A Farm in Wisconsin

A Farm in Wisconsin

Richard Quinney

Published by Borderland Books, Madison, WI
www.borderlandbooks.net

Publisher's Cataloging-In-Publication Data

Quinney, Richard.

  A farm in Wisconsin / Richard Quinney.—1st ed.

   p. : ill., geneal. tables ; cm.

  Revision and expansion of: Of time and place : a farm in Wisconsin. ©2006.

  Includes bibliographical references.

  ISBN: 978-0-9835174-0-5

  1. Quinney, Richard—Family. 2. Family farms—Wisconsin—Walworth County—History. 3. Quinney family—Photograph collections. 4. Nature observation—Wisconsin—Walworth County. 5. Walworth County (Wis.)—History. I. Title. II. Title: Of time and place.

F587.W18 Q85 2012

977.5/043   2011932663

Printed in the United States of America
First edition
Designed by Ken Crocker
Printed by Worzalla Printing
Typeset in ITC Caslon 224
Printed on 80 lb. Utopia 2XG Matte Ivory

CONTENTS

PROLOGUE

This farm, to which I am native, I want you to know about. I will describe the farm to you in words that come as I think about and imagine the lives of those who once lived here, inhabiting these few acres of rolling hills and wetlands in southeastern Wisconsin. To anyone who will listen, I have a tale to tell.

The farm has belonged to my family for four generations. Settled by my great-grandparents fleeing the potato famine in Ireland, it is the place of my birth and early years. The house at the Old Place was torn down many years ago. The well was filled with rubble to keep anyone who might be passing by from falling in. All that remains of the house that had been built by my great-grandparents John and Bridget Quinney a few years after they arrived is the crumbling foundation. I return repeatedly to view the ruins and walk the grounds.

Even now in early spring you will hear the mating calls of frogs coming from the pond at the foot of the gently sloping hill. In winter when the pond is covered with snow, muskrat houses protrude through the ice. When spring comes again, ducks and geese nest and raise their young among the reeds and tall grasses at the edge of the pond. Red-wing blackbirds establish their territories, calling from the tops of cattails. Sandhill cranes build their nests at the edge of the marsh. Prairie grasses and flowering plants and oaks and hickories grow on the hills that rise beyond the Old Place.

When we were young, growing up in the thirties and forties on the farm, my brother and I would look across the field to the Old Place and wonder about the lives of the generations that preceded us. With the passing of years, as we moved away and lived our lives in other places, the magic of the Old Place only increased in our imaginations. In times of need, as well as in times of ease, I would return to the Old Place to find solace and renewal. I go there regularly now to know that I am an intimate part of the place that is still my home, and to be reminded of the ancestors who have gone before me.

Among the things that were carried up to the farmhouse when our great-aunt Kate died in 1942 were the albums of family photographs. The sons and daughters of Bridget and John, and their sons and daughters, made the photographs as they documented their lives at the Old Place and at the farm. A few letters, some diary entries, and a scrapbook of obituaries survive for the historical record. In addition, there are the photographs in albums from my mother's family. And there are the many photographs from albums made by my mother and father from their young lives, and there are the photographs of their early years together after they married. With my birth in the mid-thirties, followed by that of my brother, the years of the growing family were well documented and preserved in family albums.

We, my brother and I, are now the keepers of the farm. Gradually we are converting the farm into sustainable agriculture. Our hope is to make the whole farm into a natural habitat. In the meantime, I have photographed the remains of the farm from earlier times. With camera in hand, I made my way to the barn, to the machine shed, to the chicken house, and to the farmhouse, photographing the artifacts of the life that once was here. These photographs pay homage to the ancestors that made this farm. The past becomes a part of the living present. Lifetimes are burning in every moment.

This farm in Wisconsin that I continue to write about was made by emigrating ancestors, by ancestors who were born here, and by those who left the place—by all those generations who have been shaped by these few acres of earth and home. These are the ancestors that I have remembered, telling the stories that have been passed to me, stories in words and images that document their lives. There is enough here to know that we are an intimate part of those who have come before us.

Our lives continue—in body, mind, and spirit—from their lives. In this real sense, there is no birth and no death. There is only the one river of life that keeps flowing, and we all are part of it. We know that our ancestors are not merely of a former time. They are with us always, in our daily lives, just as we will be in the lives of those who come after us.

THE OLD PLACE

It must have been a morning early in spring, after the melting of winter's snow and the drying of the land by the warming sun, that I formed what would become my first memory. I would have been nearly five years old. Standing beside my father east of the barn, looking to the southeast, we watched as my grandfather, nearing the end of his life, slowly made his way across the field. A dark figure coming toward us from an old place, this I would remember for the rest of my life.

I have no recollection of my grandfather's arrival at the farm after his perilous journey across the hilly field. He would die within the year, just before the war began. In one of the tattered family albums, my grandfather, John Quinney, appears as a tall, mustached man in overalls standing behind the team of horses and the grain drill in the same field where I would see him later at the end of his life. From my earliest memory to the present moment, my world begins and ends at the Old Place.

Grandfather John Quinney in the field in the 1920s

My father had been born at the Old Place as the new century began. His would be the last generation to live in the old house. A year after the last aging member of the household, my great-aunt Kate, died in 1942, the house would be torn down. As a boy I would walk across the field and look into the cavity that once was the basement. The ruins of the foundation outlined the house that once rose above and served as a home for three generations of my family.

I don't know when the house was built. It was some years after my great-grandfather, also named John Quinney, and my great-grandmother Bridget O'Keefe purchased the few acres in 1868. They had come to the United States on emigration ships from Ireland during the potato famine of the 1840s. They settled in Yonkers, New York, and in 1850 they were married. Whether John and Bridget knew each other before emigrating from Ireland is not known in our family history. We know that in 1859 they moved to Walworth County, Wisconsin. After renting an acre or two of land south of Millard, in Sugar Creek Township, they bought the land four miles south that would be the beginning of the farm.

They lived in a small house near the side of the road—where lilac bushes still grow—until building the large frame house a few yards to the south and across the road. The new location overlooked a marsh and a pond. The road curved sharply at the bottom of the hill. The center portion of the house was two storied. On either side were the wings, extending north and south, with an open porch facing east down to the pond. I remember being inside only once; my dad showed me the stairway that he had climbed as a boy each night with a kerosene lamp in hand. In a photograph in one of the old albums, Bridget, in old age, sits in a wicker chair on the sunny side of the house, hollyhocks in bloom. Photographs from the early 1900s show scattered outbuildings, an orchard, a well, and the oaks and maples that reached high into the sky.

Of the next generation, the five children of John and Bridget, some were born in Yonkers and some on the acres south of Millard, before the family moved to the place beside the road that is now called Quinney Road. All grew up at the Old Place and took their turn in the new world. Katherine (Kate), the eldest, lived in Chicago part of her life, working as a seamstress and milliner in the house of a wealthy woman named Mrs.

The house at the Old Place

Great-grandmother Bridget Quinney

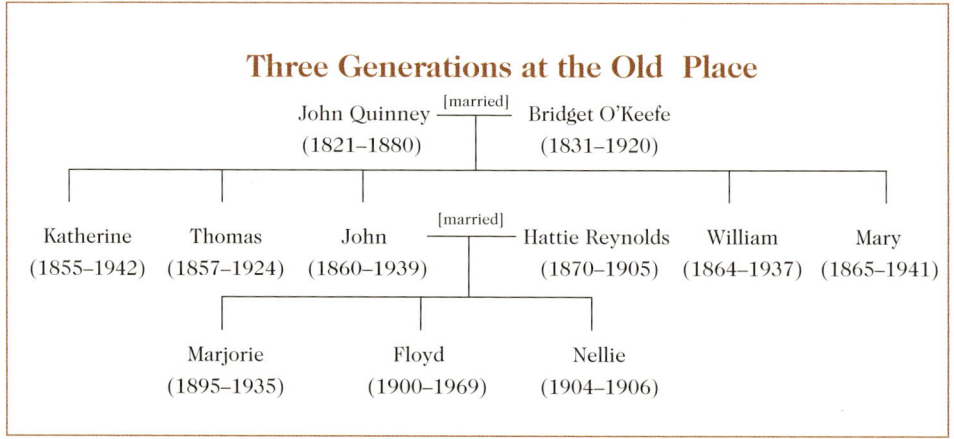

Three Generations at the Old Place

John Quinney —[married]— Bridget O'Keefe
(1821–1880)   (1831–1920)

Katherine | Thomas | John —[married]— Hattie Reynolds | William | Mary
(1855–1942) | (1857–1924) | (1860–1939) | (1870–1905) | (1864–1937) | (1865–1941)

Marjorie | Floyd | Nellie
(1895–1935) | (1900–1969) | (1904–1906)

Woodward. Thomas and William each married and homesteaded in South
Dakota, remaining in or near Alexandria for the rest of their lives. Mary,
the youngest, married Henry Reynolds and lived on a farm near Lake
Como. John, my grandfather, remained at home and bought more acres
and passed the farm on to my father.

My grandfather John married Hattie Reynolds from Rock County in
1894. Then began the third generation to live at the Old Place. Marjorie
was born first, worked as a maid and owned a roadside tavern in her adult
life, and died of a ruptured appendix at the age of forty. Floyd, my father,
was born in 1900, farmed the land, married Alice Marie Holloway, who
would become my mother, and died in 1969. Little Nellie, born in 1904,
died before she reached the age of two. This was the last generation at the
Old Place before the house was torn down.

My father and mother built a new house a quarter of a mile to the west
where the barn and sheds were located. It is a bungalow-style house, perhaps inspired by my father's trip to California when he was twenty-four.
The house was built in the summer of 1930, just in time for their September wedding. I was born in May of 1934, and given the name Earl Richard.
My brother, Ralph, was born two years later. The farm was now our home

place. Still, it was the Old Place that held all the mystery we ever needed to know. I sometimes sit on a cement bench near the empty corncrib east of the barn, looking southeast over the sloping field, and gaze longingly at the Old Place.

The lilacs and the silver maples, with new shoots and saplings, grow along the old driveway where the house once stood. Rusting fence posts and wire enclose the caved-in well that supplied the needs of three generations. A few apples hang from the branches of what is left of the apple trees. The road at the bottom of the hill is grown over with shrubs and willows. At the pond, frogs croak and chirp each summer night, and in winter muskrat houses rise through the ice and snow. Daylilies grow from the banks of the road that now passes along the south side of the Old Place. For some fleeting moments, I imagine the lives of the ones that once lived here.

Photographs in the family albums provide ample documentation of the Old Place. On a winter's day, in one of the photographs, Kate stands in the road at the Old Place and looks toward the farm. Another photograph is of Marjorie, in a checkered coat with fur stole, posing behind the Model T Ford, outbuildings in the background. Floyd is pictured in the field south of the house tending to the chickens and cows. Horses and a black and white dog appear in several of the photographs.

There is a photograph from Kate's last year at the Old Place. My father has taken his two sons, Earl and Ralph, down to the house to visit Kate. A very old Kate stands with her sister, Mary, and visitors from Lake Como. My father is in the foreground as my brother and I pose with Kate. One night, a year later in 1942, my father came back to the farmhouse and

Great-aunt Kate Quinney

*Marjorie
Quinney*

*Floyd
Quinney*

13

Floyd, Kate, Earl, and Ralph at the Old Place in 1941

14

told us that Kate had passed away. This is my first memory of a death in the family.

Life had finally come to an end at the Old Place. When the house was torn down the next year, my father lamented that Kate's box of buttons had been lost in the destruction. The lament continued the rest of his life.

I have located the original deed and mortgage, dated February 6, 1868, for the purchase of the Old Place. They have been passed from generation to generation, stored in a succession of envelopes and safety boxes. The originals are among the deeds that record the addition of acres that make up the farm. The documents describe in legal terms the specific location of the thirty acres purchased by John Quinney. The cost of the land is a sum of $300, to be paid in four equal annual payments, with an annual interest of seven percent. No mention is made of a house existing on the property. John and Bridget must have lived in a small house on the property that would serve the family for the first few years.

The mortgage papers contain the signature of my great-grandfather. The lawyer has signed the name John Quinney. On the line is a large X with the lawyer's notation—"his mark." Oh pioneers!

We are famine Irish. This fact was only vaguely recognized as I was growing up. We repeated the litany that John and Bridget had come from County Kilkenny many years ago. That Bridget had smoked a clay pipe. That Kate and Marjorie had worked as seamstresses and maids in the houses and resorts of the rich in Chicago and around Delavan Lake and Lake Geneva. Still, any details about the fleeing of Ireland in the 1840s during the height of the famine were not passed down to us. I have little idea of the difficulties faced by the three generations before me as they assimilated into the culture of the new world. Being famine Irish would have been a disgrace and a stigma on the generations of immigrants. By mid-twentieth century, one could have a sense of pride in being Irish, but one did not connect being Irish with the flight from the great famine in Ireland.

There is one stark piece of evidence that indicates to me that the generations bore a deep scar from being famine Irish. When Kate died in 1942, a few of her possessions were brought up to the farmhouse. They remained in the music cabinet, also from the old house, on the porch for half a century before I took a look at them. For many years of her life, Kate kept a scrapbook, consisting mainly of newspaper clippings of births, marriages, and deaths in the family. But midway in the scrapbook is a leaflet with a colored drawing, a derogatory cartoon about the Irish. An apelike figure with the tools of a worker poses in Irish coat and top hat. The worker has been made obsolete in the course of industrial development. Words below the figure suggest that the only thing to do is to get a job at the zoo as "some new kind of ape."

You can only guess Kate's reasons for placing the cartoon in the scrapbook. She certainly was recognizing her Irish identity—a lifetime of identification. The message that the Irish are less than human could not be avoided, likely striking a deep chord and reminding her of the cruel and unfair treatment of the Irish. Such characterization of the Irish could easily cause you to flee from a public identification of being Irish. Just as you had fled from the shame of the great famine.

Although I have tried to trace my Irish roots further back than the famine, searching the national archives in Dublin and the parish records in County Kilkenny, I have failed to learn much about my ancestors before the emigration of John and Bridget. There are bits of information about the possible spellings of "Quinney," variations due to transcriptions of

YOUR OCCUPATION'S GONE.

Your job is played out, for they need no more Micks ;
Steam-engines now hoist up the mortar and bricks :
Your prospects, I think, are exceedingly blue,
And the only thing left you, in this pinch, to do,
Is to try, on the strength of your mug and your shape,
For a job at the Zoo as some new kind of ape.

Cartoon of the Irish

the Gaelic, and a possible change in spelling during an early migration from northern to southern Ireland. Quinney may be an Anglicized form of the Gaelic name O Coinne, a descendant of Coinneach. One genealogist that I employed suggested that we descended from a Tyrone O'Quinney who marched with the army of Eoghan Rua O'Neill to the confederation of Kilkenny in the 1640s. The same researcher came to the following conclusion: "It is clear to me that your ancestors belonged to the cotter/labour class, which suffered most as a result of the Great Famine and generally featured in the records as mere statistics." I am assuming this conclusion is the bare fact of my Irish ancestry. A fact that subsequent generations of the family remembered uneasily, and hesitated to pass on to subsequent generations. There was a toll to pay, one that continues to my generation and perhaps beyond.

The practice of the Catholic religion had dissipated with the passing of generations. My father would tell of Bridget walking the five miles to attend mass at St. Andrew's Catholic Church in Delavan. Kate, I know, was a devout Catholic. A black glass rosary rested in a box of trinkets in the attic of the farmhouse. I never heard my father speak about religion in his own life. My brother and I were sent to Sunday school at the Delavan Methodist Church while our parents attended the morning service. Other than always trying to do the right thing, we did not talk about religion at home.

I have always wondered when the Irish accent ceased to be heard among my immigrant ancestors. We would imitate the Irish brogue on St. Patrick's Day, vaguely recalling our original tongue. Did Kate and her brothers and sister speak with an Irish accent? As part of assimilation into the dominant Anglo culture, removing the Irish from your daily speech would be an important thing to do. My guess is that tones and inflections from the old country continued to be heard at the house on the hill. Passing by the Old Place, now, I hear a sound that comes from another land and another time.

I grew up on the border, and a life has been made from this reality. Being a descendant of emigrants naturally sets the pace for a life. The emigrants, the generation that left the old country, experienced the pains of entering a new land. The descendants of emigrants, generations later, are only a few steps removed from the immigration to a new land. The old country is a faint memory passed through the generations. Assimilation of the generations into the ways of the new land draws from the past to give identity and a sense of being rooted in something from another time and place.

Our world was one not only one of emigrants but also of immigrants who were farmers. Five miles away from the farm were the small towns of Delavan and Elkhorn. These towns were of the other world—streets, brick buildings, offices of doctors and lawyers, churches, and schools. We were also in the orbits of Chicago and Milwaukee, each sixty miles away. Kate had commuted to Chicago on the train for years before the turn of the century to work in the house of the wealthy Mrs. Woodward. Marjorie worked in Chicago for some years as well. Walworth County, around Lake Geneva and Delavan Lake, had been the vacation destination of the rich and the rising middle class of Chicago and its suburbs. We of the country regularly came in contact with these strangers. We could not remain provincial when we were in the midst of others.

Occasionally we ventured to Chicago and Milwaukee to see a traveling Broadway musical, to go to a baseball game at Wrigley Field, or to shop and see a movie in an ornate theater on State Street or Wisconsin

Avenue. By train or car, the trip to the city was an adventure that would hold us from one time to the next. It is a pleasure still to remember my father dressing in his suit, putting on his topcoat and fedora, and heading to Chicago to attend the international stock exhibition. Once he brought home a toy bus for each of us.

In an essay "Something to Write Home About," Seamus Heaney describes the god the Romans called Terminus, the god of boundaries. The Romans kept an image of the god in the Temple of Jupiter on Capitoline Hill. An opening to the sky was in the roof above, as Heaney writes, "as if to say that a god of the boundaries and the borders of the earth needed to have access to the boundless, the whole unlimited height and width and depth of the heavens themselves." Even with boundaries and borders, or especially with boundaries and borders, both gods and humans need to feel that they are unbounded. "A good poem," Heaney adds, "allows you to have your feet on the ground and your head in the air simultaneously." And to this I would add, life is both earthbound and open to the spaces beyond. We are simultaneously here and elsewhere. I learned this well on the farm, and down at the Old Place.

All that I know about the life of my great-aunt Kate could be written in a few lines. Little about her has survived for the telling. One reason for the scarcity of information is the lack of people to pass on thoughts about her. Kate never married, and our families have been small in number. I have found in the top drawer of the buffet at the farm a diary that Kate kept. The diary is for the year 1892, but she used it over a period of years to record a few brief entries. An entry dated 1924 lists household items; there are a few addresses of relatives; the deaths of her brothers Tom and John are recorded, including the times of death; and she enters the births of my brother and me. In the entry for my birth, she adds, "A very good looking chap." On the opposite page, in a child's hand that must be mine, my name is scribbled, indicating that I had some contact with Kate, although I do not remember this. Midway in the diary there is a pressed four-leaf clover. We Irish believe, superstitiously, in four-leaf clovers.

Through good fortune the photograph albums that Kate kept, over a period of nearly forty years, have survived. Since discovering the albums on the front porch of the farmhouse several years ago, I have gone through them with ritualistic regularity. I am the family archaeologist who digs in the buried past to learn about that other time and about those other lives to which I am biologically and psychologically—and spiritually—connected. I am the archaeologist who searches for the artifacts that will shed light on those who came before us. The search, and whatever knowing that may result, is our way of keeping the ancestors with us.

Without the stories of those who knew Kate, and without the memories of subsequent generations, I am left to imagine Kate's life from the photographs glued to the black pages of the albums. Without words to read, or stories to remember, I have a glimpse of her life in the visual record she created, as selective as this may be. Each time I study the tattered albums, the world they portray becomes stranger and more complex to me.

The albums contain much information that will never be retrieved. Historians, archivists, and living relatives will never be able to identify many of the people and places that now exist only on paper. The woman in peasant clothing who sits in front of the shed tending her flock of ducks could be in Ireland, or South Dakota, or Wisconsin. There are photos of horses and buggies on streets in small towns somewhere. A tom turkey struts for the hens in a backyard. Waves lap the shore someplace along Lake Michigan. Institutional buildings of brick loom in the

21

distance. Studio portraits of men, women, and children—lost to our memories—gaze at me as I turn the pages. In a meadow beside a creek, the cattle are lowing, and a baby sleeps.

Kate appears often in a fine tailored suit of checkered cloth. She is pictured at the Old Place with outbuildings, marsh, and woods in the background. My dad and Marjorie are in photos at various times in their lives. Bridget is photographed beside the house, a scarf on her head. Large family gatherings are posed on the farmhouse porches. My grandfather John is shocking grain in the field. Kate sits on the edge of an imitation moon, likely from a day at Riverview Park in Chicago. Again in the clothes and the hat that she has fashioned, Kate stands on the shore of Lake Michigan; a sailing ship is docked in the harbor.

Mary, Kate's sister, appears in numerous photographs in the albums. The youngest of the five children of Bridget and John, she moved to the farm at Lake Como after marrying Henry Reynolds (no relation to Hattie Reynolds) in 1881. From then on, many of the family gatherings were held at their home and recorded in photographs taken on the front lawn. Some photographs show Mary and Henry spending afternoons at the Old Place. Their children, Howard and Birdina, are pictured at various stages of their lives. Howard's son Gene, my second cousin, and his wife, Betty, used to visit us when we were at the farm. Recently I found Mary's grave in the cemetery at East Delavan.

A motif of farm and city is evident throughout Kate's albums. That she lived part of her life in the city, while simultaneously returning to the Old Place sixty miles to the north of Chicago, is the theme of her life. Thirty-some years ago, I spent several days at the Chicago Historical Society researching the outlines of Kate's life in Chicago. Her albums contain many photos of her employer, Anna G. Woodward, widow of Morgan S. and remarried to Emmet B. Thompson, who resided at 414 East Forty-fourth

Kate in Chicago on the shore of Lake Michigan

23

Street from 1909 to 1916. With the address and Kate's photo albums, I found the building on the corner of East Forty-fourth and Martin Luther King Boulevard, once named South Parkway. The building, with its stone carvings and cast-iron fence, was just as pictured in the albums. In a photograph in one of the albums, Mrs. Woodward is seated in her stylish electric automobile.

I imagine Kate working at her sewing machine behind the upstairs window in Chicago. Some days she takes a streetcar uptown to Marshall Field's to shop for material for the making of exquisite dresses and hats. On clear days she walks east to the shore of Lake Michigan before returning to the house of Mrs. Woodward.

Kate's albums also note the high-society adventures of Mrs. Woodward. One newspaper clipping, announcing the marriage of Anna Graham Woodward to Emmet B. Thompson at the Church of Transfiguration, proudly notes: "The bride was in Peking during the Boxer uprising as a guest of Minister Conger, and became known to the diplomatic corps as the 'heroine of the Peking siege.'" Her son, Lieutenant Warren Woodward, who was in the flying service, was killed in France during the war. On another page is a photograph of Mrs. Woodward in a rickshaw, and right below it is a photo of Kate taken at the Old Place. Elsewhere, Mrs. Woodward is seated in her fancy car, parked in front of her building; right below is a photograph of Kate sitting on the running board of the Model T, crocheting, at the Old Place. It is a day during World War I, and a flag hangs in the front window of the car. At the end of the album is a postcard of the residence of Mrs. Potter Palmer, and an invitation to Mrs. Woodward to attend an afternoon party in the Shakespeare Garden. Also, there are cards of scenes from Ireland that have nothing to do with Mrs. Woodward.

From the probate records at the Walworth County courthouse, I have found that at the time of Kate's death in 1942, her personal estate was

*Mrs. Anna
Woodward
in Chicago*

*Kate
crocheting
at the Old
Place*

25

$625.37 from cash in the Wisconsin State Bank. Her household furniture was valued at $25.00. She had one-fifth interest in the property and house at the Old Place. I keep Kate's writing desk, a "secretary," in my living room in town. Someday I will donate her photograph albums to the state archives for safekeeping. Her inkwell, made of iron and green glass, is on my desk, providing another day of remembrance and inspiration.

Kate kept a separate album of newspaper clippings. The cover has the color of a faded rose with a binding that is in threads. I leaf through the album to know more about the events, from birth to death, in the lives of my ancestors. I learn about the things that mattered to Kate.

There are the obituaries clipped from local newspapers and the obituaries sent to Kate from South Dakota. Bridget's obituary of 1920 is pasted directly to the inside of the front cover. Her grown children survive her: John at the farm, William and Thomas in South Dakota, Mary Reynolds on the farm at Lake Como, and Kate in Chicago. I had not known that a child had died in infancy shortly after her emigration from Ireland and marriage in Yonkers. Beside Bridget's obituary, Kate pasted a poem with lines that read:

> I did not weep.
> It seemed I did not know
> 'Twas endless sleep.
> And time went on—
> Drab days that groped or sped.
> Somehow I could not learn
> That she was dead.

26

There are the clippings of the weddings, the births, and the anniversaries. But the reason for the making of the album of clippings is the fact of death, the Grim Reaper. The album is writ large as a cautionary tale that reads *memento mori:* you must die. The obituaries accumulate and accelerate in number. Brother Tom dies in December of 1924; niece Marjorie dies in October of 1935; brother Bill dies in January of 1937; brother John dies in January of 1939; and sister Mary dies in September of 1941, only a few months before Kate's death. In between are the deaths of young nephews and nieces from infectious diseases and accidents on the highway. From the obituaries I learn about the lives of those who once lived and were remembered in death, in the clippings brittle and fast fading.

Kate followed the fortunes of the hymn "Sweet By and By," which was written in 1865 in Elkhorn, just seven miles east of the Old Place. From the *Chicago Tribune,* Kate pasted the lyrics by S. Fillmore Bennett, with the refrain:

In the sweet by-and-by,
We shall meet on the beautiful shore.
In the sweet by-and-by,
We shall meet on the beautiful shore.

Later in the album is a clipping about the settlement in an equity court in Boston of the legal rights to the royalties of the hymn. When Joseph P. Webster wrote the music for the hymn, he had signed a contract with Lyon & Healy in Chicago to publish the hymn. After the Chicago fire of 1871, the rights were sold to the Oliver Ditson Company in Boston. The lawsuit in the equity court returned the royalty rights to the Webster estate. Kate's clipping describes the origin of the hymn:

It was written, words and music, in less than an hour and was inspired by a temporary fit of depression of Mr. Webster. It is related that in 1865 Mr. Webster went into the home of his friend, Dr. Samuel F. Bennett, in a most despondent mood. The doctor asked what was the trouble. "It is no matter, it will be all right by and by," was the reply. This remark acted as a flash of inspiration to them both. Dr. Bennett immediately sat down and wrote out the verses and Webster composed the music on his violin. Less than an hour later they were singing the song with two friends.

Joseph Webster continued to teach music in Elkhorn, and wrote the song "Lorena," which became famous after the Civil War. Like many other girls of the same time, my mother's mother was named Lorena.

In an article dated March 17, 1937, pasted into Kate's album, President Franklin D. Roosevelt lauds the spirit of Saint Patrick. On that day, Roosevelt, delivering a message on the 125th anniversary of the Hibernian Society of Savannah, Georgia, and the 200th anniversary of the Charitable Irish Society of Boston, decried that selfishness was the greatest danger confronting the nation, and urged Americans to follow the footsteps of Saint Patrick and his epitome of unselfishness. The motto of the Irish societies—"not for ourselves but for others"—may well be "the inspiration for all of us," the President said. Four years later, Kate pasted into her album the article reporting the burial of the President's mother, Sara Delano Roosevelt. To the end, Kate's heart was with the Irish and those who supported them.

I sit here at my desk carefully holding the scissors that Kate used throughout her life of work as a seamstress. The scissors are heavy and large, ten inches in length, and the handles are black. I use the scissors occasionally to cut pages from newspapers of the day. As you may guess, the material things gathered from the Old Place, and the lives imagined of those who once lived there, are the Muses of my life. I need not look farther than the Old Place; it has been my inspiration for a lifetime.

Kate would occasionally visit Tom and Bill and their families in South Dakota. In the trunk on the front porch I found a postcard Kate received from Tom. On the front of the card, Tom and his wife, Florence, are standing in front of their house. On the other side of the card, Tom asks Kate to come out to South Dakota for a visit. A few years later, in the early 1920s,

Florence and Tom Quinney in South Dakota

Kate makes the trip. A photograph in her album shows Kate and her brothers on a summer's day with the trees sending shadows across the lawn. They are pleased to be together as they pose for the camera.

About all I know of Tom and Bill is gleaned from the newspaper clippings Kate placed in her album. Families pass stories about their uncles and aunts, and great-uncles and great-aunts, from one generation to another. I am fortunate to have the obituaries that Kate saved, perhaps with other generations in mind. Obituaries published in newspapers for the public to read are the documents for this family's history.

The headline reads, "Thomas H. Quinney Called by Death." A pioneer of Hanson County, he died at the age of sixty-seven at his home in Alexandria. Tom had traveled to the Dakotas in 1880 to take up a homesteading claim. A year later, he returned to the boyhood farm in Wisconsin and claimed Miss Florence Loomer of Millard as his bride, who survives him. Also surviving is a son, Elwin, assistant engineer for the State of South Dakota. A daughter, Lillie May, died in 1919 of influenza and pneumonia.

Tom's travel to the Dakotas in 1880 to claim a homestead has been described in an archive of the lives of prominent old settlers of South Dakota. Starting from Whitewater, a few miles north and west of the Old Place, on February 15th, Tom and his friend C. H. Nott traveled by rail to Algona, Iowa. Caught in a snowstorm, the train and passengers were blockaded for four days. After shoveling snow, they pushed on to Emmetsburg. Tom developed back trouble and was forced to lie quietly in a hotel for two weeks. Pressing on by train, the travelers were again blocked by snow, on their run to Mitchell. Tom and his friend Nott were now discouraged of reaching their destination by rail and, finding their funds running low, decided to walk. Pushing out across the snow-covered prairie, with the white unbroken expanse reflecting painfully the sunlight, they were snow blind by the time they reached Marion Junction. Somehow, the men

Tom, Kate, and Bill Quinney in South Dakota

continued on the next morning, groping in darkness in a strange country and without even so much as a trail to follow.

As they struggled on, sustained by hardened snow crust, but suddenly sinking where the tall grass had weakened the crust, Tom and his companion stumbled up the hills and plunged through the ravines, until finally they heard, from a distance, a man calling his cattle. By calls and answers, they succeeded in reaching him and passed the night in his shanty. The next morning, not yet recovered from their snow blindness, they hired the man to lead them to the Barker brothers, who had been old friends of theirs in Wisconsin, and who lived but a few miles away.

As soon as Tom and Nott recovered from the effects of their walk in the snow, they went to Mitchell and filed their homestead claims. After purchasing three dollars' worth of roof boards for their sod shanty, a sack of flour, and a jug of molasses, they took inventory of their cash and found they had just fourteen cents. But in time Tom obtained a fine property and a comfortable fortune.

This from Tom's obituary:

> Death came to Mr. Quinney in the early morning of December 16. He had been ill for several weeks, and for the past two weeks or more it was known by the relatives and friends that his passing was but the question of a short time. He made a valiant struggle against the Grim Reaper, but it was a vain effort, and his never strong constitution was not equal to the struggle.

And we are reminded of his accomplishments:

> Mr. Quinney was well known and was held in high esteem by a large circle of friends in the county. He served for a number of years as commissioner of Hanson County, having also served as chairman of the board. For a number

of years he was president of the Hanson County Agricultural Society, and always had a great interest in the county fair. He was a member of the Alexandria Odd Fellows lodge, and was also affiliated with the local lodge Ancient Order of United Workmen. Mr. Quinney was a good neighbor, a loyal friend, respected by all who knew him.

Tom and his family had lived on the farm eight miles south of Alexandria until they moved to town in 1910. Many friends and neighbors were present at the burial at the Green Hill cemetery as the Odd Fellows ritual was conducted.

There is a photograph of Aunt Florence, Tom's wife, in our family album. She is kneeling on the lawn in front of our farmhouse with her arms around my brother and me. I know that she attended my grandfather's funeral in 1939, her name being in the visitation book. I keep on a shelf of my living room bookcase the copy of nursery rhymes that she gave to me in 1941 when I was seven, with her inscription "From Aunt Florence Quinney." I remember my father and mother telling me that Florence had told them that one of the great hardships for her when she and Tom were homesteading was keeping rooms clean in the sod house.

A clipping from the Alexandria newspaper describes a surprise party for Tom and Florence in 1921 at the time of their fortieth wedding anniversary. The title of the article reads "Mr. and Mrs. T. H. Quinney Victims of Surprise."

Wednesday of last week was the fortieth wedding anniversary of Mr. and Mrs. T. H. Quinney, and as they were among the real pioneers of Beulah Township, several of their old neighbors of the early days planned a surprise upon the worthy couple. Mr. and Mrs. Quinney were invited to the home of Clifford Shade for supper, and while they were gone the friends made arrangements for the party, and gathered at the home early in the evening, taking care to see that no lights were going about the time the victims were expected to arrive home. The house was filled with old neighbors and friends to the number of seventy when Mr. Quinney opened the doors, and when he switched on the lights his face was an index to the great surprise with which he was confronted. He could not say a word for some time, but finally recovered enough to ask if there was anybody in the cellar.

The evening was spent in singing old time songs and talking over the old times down in Beulah Township, in which reminiscing nearly every one of the guests could join. The "gang" had come prepared to enjoy the entire evening, and naturally had not forgotten the "eats," which were excellent and plentiful. It was a late hour when the gathering broke up after presenting Mr. and Mrs. Quinney with a handsome electric table lamp.

Tom has not quite recovered his composure yet, and both he and his estimable wife are still wondering where all the people came from, but both are unanimous in the opinion that the event was a complete surprise, and that they had a mighty enjoyable time.

Tom and Florence had two children. Their son, Elwin (married to Juliet Mosier), became an engineer for the State of South Dakota. Only recently have I had the good fortune of making contact with his descendants. Tom and Florence's daughter, Lillie May, taught in the district schools of Hanson County until the spring of 1916 when she began to suffer from pneumonia. Lillie died of pneumonia and influenza three years later in Albuquerque, New Mexico, where she and her husband, Frank Rollings, had gone to seek relief for Lillie's health. Frank died within two days of Lillie's passing as the train was returning to South Dakota. The obituary, captioned "Fond Hearts Are United in Death," reports and describes the double funeral.

William H. Quinney, Tom's brother Bill, moved to Hanson County in 1887. He was united in marriage the same year to Miss Agnes Bamber. Bill and Agnes homesteaded for some years, and then moved to Alexandria where Bill served as chief of police and night watch for many years. He died in 1937, and the obituary tells us that he "was possessed of a friendly disposition, and was well liked by all who knew him. He was accommodating to everybody, and was willing at any time to extend aid to those in need to the extent of his ability. As one of the 'old timers' in Alexandria, 'Bill' Quinney will be missed and sincerely mourned by many friends." He is buried in the Catholic cemetery in Alexandria.

In the album, we learn that Agnes, who died sixteen years before Bill of appendicitis and a heart attack, was born in Milwaukee and came to Hanson County with her family in the early 1880s, right after the railroad was built through Alexandria. Her family settled on a claim south of Alexandria, where she lived until her marriage to Bill in 1887. The obituary of 1921 tells us that "coming from a city to this new life was a great change, but Mrs. Quinney easily adapted herself to pioneer life, and during her later years she took pride in the fact that she was an early settler of Hanson County." The obituary continues: "Mrs. Quinney's life and happiness centered about her family and her home, where she was perfectly happy in contributing to the comfort of her loved ones and her friends. She was of a kindly disposition, friendly towards everyone, and was of a hospitable nature, enjoying the companionship of her friends and neighbors. She was patient, cheerful, and always ready to give help where it was needed."

A postcard in Kate's album illuminates the page. Printed in Saxony, by appointment of the King and Queen, it pictures the cliffs of Dover jutting into the sea. The card is dated October 1908, Alexandria. In a beautiful hand, the card is addressed to W. H. Quinney, Delavan, Wisconsin, c/o John Quinney. From the note written on the card, we know that Bill has gone back to the Old Place to visit his ailing mother, Bridget. Agnes writes to Bill:

> Your postal received. Glad to hear Mother is better. Mary is still very sick so I wish you would come home. You did not acknowledge my letters. Did you not hear from us since you have been there? It seems an age since you left. I hope Mother will get entirely well. But come home as soon as you can. Yours always, Agnes

Agnes would live for another thirteen years, and Bill would live another twenty-eight years. "Come home as soon as you can"— Agnes's invitation and plea. It is a call that both beckons us and cautions us to treasure the moments we have now.

The lives and fates of the five children of Bill and Agnes are part of the history of the first generations of homesteaders and pioneers on the frontier. The tales told in the newspaper clippings and obituaries readily convey the sorrows the families knew in Hanson County and in the small town of Alexandria. There were four daughters, all born in the decade of the 1890s: Angela (married John Wagner), Mary (George Poland), Pearl (Henry Johnson), and Louella (Clarence Boos). Bill and Agnes adopted their grandson William ("Willie") after the deaths of his parents, Angela and John. Angela died when Willie was two months old; John died two years later. At the age of nineteen, in 1929, his first year of college, Willie was killed with three other friends in an automobile accident in Sioux Falls.

The lives and the deaths of two of the daughters at a young age, and the death of the grandson, are documented in Kate's scrapbook. The sorrowful writing in the obituaries is not far removed from our own sorrow and

36

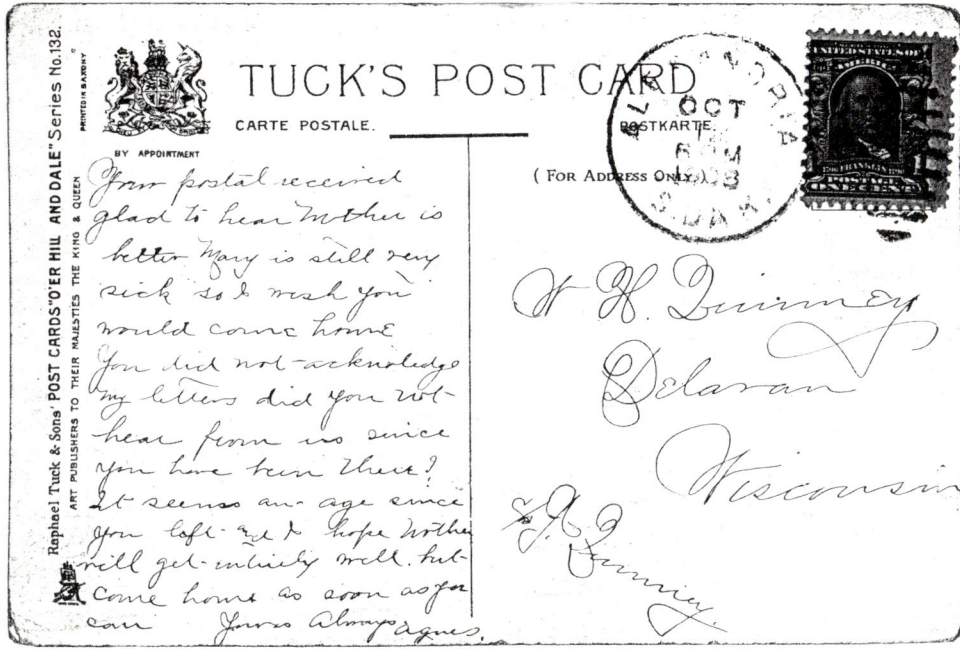

Postcard Agnes Quinney sent to Bill in 1908

bewilderment when death comes. Kate was our keeper of family history and family grief. We must imagine there were good times in daily life that escaped the telling. Our family awaits its novelist.

Hattie, my father's mother, the grandmother who died twenty-nine years before I was born, I knew only as the image in a large oval portrait labeled "Floyd's Mother" in the trunk on the front porch. My father told me the story several times that when she died of tuberculosis at the age of thirty-five in 1905, his father said that he would never marry again because he would never find a woman as good as Hattie.

Hattie's father and mother were Nathan Church Reynolds and Mary Potter Reynolds, who married in 1851 in Rock County, Wisconsin. Nathan, a son of Moses Reynolds, had immigrated to Wisconsin from Waldo County in Maine, and Mary had emigrated from Allegany County in New York. Nine children were born to Nathan and Mary Reynolds, and grew up on a farm near Milton. From genealogical research, we know that Nathan descended from a family that dates back to the Puritans of Salem and Plymouth County in Massachusetts. Family genealogy traces the family back to Sarah Pease, who was accused of witchcraft in Salem in 1692. Daniel Reynolds, an ancestor, possibly directly related, fought in the American Revolution.

The stories of the children of Nathan and Mary Reynolds begin with their first child, Seymour. Seymour loved horses and raised colts with care and affection. When Hattie died, and John moved across the road to live in Bridget's house, Seymour, always a bachelor, moved into the small house by the lilac bushes. He liked to play cards, and in the winter of 1913, he and some neighbors played until early one morning. After the neighbors had left, Seymour found that there was no wood in the house

Mary Potter Reynolds   *Nathan Church Reynolds*

to keep the fire going. He walked to the woods, across the snow-covered field, and fetched a log that he began to carry back to the house. He fell as he stepped on a patch of ice that was covered with snow. He was struck by the log and died. The dog that had accompanied him lay faithfully beside the body for three days, melting the snow to the ground.

Ellen Reynolds was born in 1855. She married William Webb and had four children. She became ill with cancer of the breast, had an operation in Milwaukee, suffered a relapse, and died. Her brother John Reynolds, like his brother Seymour, also loved horses and would bathe them after a day's work in the fields. After a marriage that ended in divorce, he lived in various places and eventually contracted tuberculosis and died in 1907. It is told that his last request was for a stick of candy.

Lucy Reynolds married a man named Ed Dyer, and they moved to Chicago where he would work in the construction of the World's Columbian Exposition of 1893. Lucy moved back to Walworth County after her husband died, and married a second time.

Anna Reynolds created a long lineage that covers over half of the Reynolds family genealogy. She and her husband, Merritt Baker, moved to Oklahoma where land was cheap. She moved back to Walworth County after her husband died, remarried, and in her final years tried to make a living crocheting rugs to support herself after her second husband, Dwight Wheelock, died.

Nathan Reynolds, Jr., remained unmarried throughout his life. He suffered from spinal meningitis when young, which left him deaf and unable to talk. The story is told that when he was seven he was struck by lightning while standing by the chimney in the kitchen. In his later years, he was employed as a kitchen helper in the Elkhorn Hospital, and that is where he died.

Alice Mary Reynolds was born in 1869 near Milton. She married Arthur Wheelock from Milton Junction, who would work for the Northwestern and Milwaukee railroads, and then for the Cudahy Brothers Packing Company in Milwaukee. Alice eventually moved to Whitewater, after a divorce, lived alone, and supported herself by taking in washing and ironing. She was finally cared for by a granddaughter, and died in 1958. She was the mother of eight children.

Hattie Reynolds, my grandmother, was born in 1870 on the farm near Milton. Little information is provided about her life in the family genealogy. I know that she moved to the house at the Old Place when she married John Quinney in May of 1894. When she died, her baby, Nellie, was one year old; Floyd was five and Marjorie ten. I take flowers to Hattie's grave at Spring Grove Cemetery in Delavan every spring.

Nellie Reynolds, born in 1873, was the youngest of the brothers and sisters. She was engaged to marry a man of whom her parents did not approve. Her sister Anna had made a blue brocaded silk dress for her wedding, but Nellie died less than two weeks before her seventeenth birthday, and instead of being married in the dress, she was buried in it. Nellie had worked for a woman with six children, and during that time contracted the tuberculosis that caused her death.

The six Reynolds sisters appear together in a formal studio portrait. My grandmother, Hattie, is standing in the back row, far left. We of the subsequent generations gaze into the eyes of our ancestors, and hope to know things beyond the few words that have survived.

I have found another studio portrait, taken in 1900 in Alexandria, South Dakota, that includes Hattie. In that year, Hattie and John and their children Marjorie and Floyd made the trip to South Dakota, with Bridget, to visit Tom and Florence and their two children. Bridget, in her seventieth year, is in the center of the portrait. John and Hattie are on the right, and the child who would become my father is on his father's knee.

After Hattie died, the girls were raised a few miles away in the home of their aunt Mary Reynolds and uncle Henry on the farm near Lake Como. Floyd stayed at the Old Place and was raised by his father, John, and his aunt Kate. Little information has passed to me about Marjorie's life from her tenth year to adulthood. And the rest of her life is a mystery to those of my generation. Whenever I would ask someone who might have known Marjorie or heard about her, the conversation would immediately stop. Or, at most, I would receive an answer such as "Oh, there isn't much to say," or "We didn't know her very well."

Portrait of the Reynolds sisters. Front row, left to right, Anna, Ellen, and Lucy. Back row, left to right, Hattie, Alice, and Nellie

Portrait of Bridget with the families of Tom and John Quinney in 1900

I turn to the photo album that Marjorie left when she died in 1935 at the age of forty. Photographs that she took and others that she collected may provide the only remaining window to her life. They allow us, with some imagination, to know about the life that would end abruptly, too soon, from the peritonitis that followed a ruptured appendix.

For years I have kept on the top of my bookcase a framed studio portrait of Marjorie. Her face in soft focus, a string of pearls gracefully around her neck, she poses in a dark dress with a delicate lace collar. She looks away from the camera, ever so slightly, lips softly closed in a slight smile.

Fading photographs glued to the black and now brittle pages make up the album that Marjorie kept. The album documents the years from about 1900 to the early 1930s. There is a photo of Marjorie and my father, labeled "Floyd & Myself," that must have been taken by a traveling photographer. My father appears to be about three and Marjorie about eight. My father is dressed in a skirt and a ruffled blouse, a conventional outfit for an Irish boy, beside a toy wagon. Marjorie wears a stylish long-sleeved dress and rests her arm on a kitchen chair. They are standing in front of the lilac bush. Twenty years later they pose for a photograph taken in a studio in town.

Floyd and Marjorie at the Old Place, 1903

In the album are the photos that Marjorie took of her uncle Tom and aunt Florence on a visit to South Dakota. Many others show family gatherings at the Old Place and at the Lake Como farm of Henry and Mary Reynolds. I pause for a long time to study the photograph of Marjorie in the front yard of the Reynolds farm. Marjorie stands between Uncle Henry and his son, Howard, who is posing with two horses. A dog keeps watch.

I also find the only pictures of my grandfather John as a farmer. He is working in the fields with his horses. The photos are labeled "Dad." There are a few photographs of Bridget near the end of her life. And there is Kate dressed in fine dresses at the Old Place and in front of mansions in Chicago and on the shores of Lake Michigan. In another photo, taken from the passenger side of the Model T, the hood ornament rises from the lower

Front yard at the Henry and Mary Reynolds farm, with Howard, Marjorie, and Henry

On a road near Delavan, 1920s

corner of the picture as the riders travel down a tree-lined gravel road. I study the buildings and landscapes in the backgrounds of the photographs, and identify some of the places that I continue to see and to think about.

People and places unidentified in photographs will remain unknown to future generations: soldiers on leave during World War I, people at picnics beside lakes and rivers, young men and women posing and frolicking in bathing suits. Horses and ponies hitched to carts and wagons, motorcycles in driveways, a biplane parked in a farm field, and dogs running across a snow-covered pond. One could get lost in other times and places.

In Marjorie's album are the pages containing the photographs of the man she knew late in her life. His business card is attached: "Lloyd L. Latta, Auto Repairing and Electric Work." The large house of his family, on Highland Avenue in Clinton, appears prominently in pictures of Marjorie and Lloyd posing in various states of composure. A few years

46

Swimming at Delavan Lake

ago I knocked on the door of the house and was given a tour by the present owner. I learned that Lloyd's sister Ruby, also pictured in the album, was still alive and lived at the top of the street. She died later, while I was gathering the courage to ask the questions about Lloyd and Marjorie.

The silence that inevitably followed any of the inquiries I made about Marjorie no doubt came from the fact that she was a single woman who for the last part of her life owned and operated a tavern. While other women in the area during the twenties and thirties were marrying farmers or teaching in country schools, Marjorie was taking another direction. Her life aroused in others both fascination and condemnation. Marjorie has been the woman of mystery all my life.

Before operating the Shingle Inn, a tavern located on Highway 14, five miles south of Delavan, Marjorie worked as a maid at the Delavan Inlet Inn and at the Wisconsin School for the Deaf in Delavan. I assume that she lived in the house at the Old Place with her father and Aunt Kate a good part of the time during those years. The Shingle Inn, which she owned in partnership with Joseph McCabe, had served as a bootlegging operation during Prohibition. It exists today (under another name) as a "gentlemen's club" featuring dancing girls, and I pause each time I pass by.

Years ago I went to the Walworth County courthouse in Elkhorn to look at the probate records that were filed after Marjorie's death on October 31, 1935. The heirs to the estate were John Quinney and Floyd Quinney. The personal estate was listed: Cash, State Bank of Elkhorn ($351.50); Cash, Citizen's Bank of Delavan ($41.73); Trust Certificate, Citizen's Bank ($187.83); Cash on hand ($35.00); and Ford Coupe, 1930 model ($60.00). Marjorie's one-half portion of the Shingle Inn at the time of her death was noted, along with the value of various items, including liquor, tobacco and cigarettes, candy and nuts, glassware, two tables and six chairs, radio, cabin, three beds and springs, and a cash register.

Periodically—out of need it seems—I page through the photograph albums that survived the destruction of the house at the Old Place for glimpses of my father's life when he was a young man. Except for the childhood photograph of him with Marjorie beside the lilac bush, the photos begin when he is in his teens. Sometimes he is pictured in the work clothes of a farmer, and other times he is dressed in a suit, shirt, and tie and wears a fedora. He is pictured with aunts and uncles, with

Marjorie at the Old Place

other people of his age engaged in various activities, and with his Model
T Ford. I favor, especially, the photographs of him working with horses in
the field or hauling cans of milk to the factory in town. After finishing the
eighth grade, rather than going to high school, he stayed at home to help
his father with the farm work.

Two items from my father's life I keep close to me. One is the small oil
lamp that he carried nightly as a boy up to his bedroom in the house at
the Old Place. He often told me about going up the dark and narrow stairway with the light of the lamp. I keep the lamp on the high shelf of the
corner cabinet in my house.

The other artifact—from the archaeology of my father's life—is the
camera he used to take many of the fine photos that appear in the family

albums. The camera, a Kodak Autographic, has a bellows that folds out and is held at the waist. I keep it in the middle drawer of the chest beside my writing desk. This is the camera that he took on his trip to California in 1924.

Several years ago I found in the music cabinet on the front porch of the farmhouse the cards and letters that my father had written to Kate and Marjorie, along with the negatives for the photographs he had taken. They provide details of his cross-country trip in the Model T Ford with his good friend Mervin Kittleson. Parts of the Lincoln Highway had been completed, and other parts were in various stages of construction.

Mervin Kittleson and Floyd beginning the trip to California, 1924

With the fall harvest completed, the travelers departed from the driveway of the Kittleson farm on the morning of September 16, 1924. By
nightfall they had reached Clinton, Iowa. My father immediately wrote
a postcard to Kate: "Arrived here at Clinton, Iowa at six o'clock. Are
spending the night at the tourist camp. Have just gone over the Missis-
sippi River. Had three flat tires, but didn't have to buy any new ones."
Five days later, upon arriving in Laramie, Wyoming, he writes to Marjorie about the hills and the boulders, the pine trees, and the snow on the
mountains. He concludes the letter: "We meet cars from every state on
the Lincoln Highway. Lots of tourists at the tourist camp. We use our pil-
lows twenty-four hours each day. Sleep on them at night. Sit on them all
day. We intend to get to Salt Lake City in about three days but you won't
have time to write us there. And we don't know which trail we will take
from there. But in case of sickness or anything like that a telegram would
find us at the tourist camp. Had a chance to get a job with a threshing
gang. Will close and make our bed. As ever, your brother."

They arrive in San Francisco on October 3. Writing to Kate, my father
notes that thus far the cost for the trip is $52 each, including all auto
expenses, eats, groceries, and camp fees. "A slow rain all day. Jobs seem
scarce around this city. If we don't find work here we will start for Los
Angeles." Later in the week they arrive in Los Angeles, and my father
writes to Marjorie: "Got to Los Angeles at noon. Reminds me of Chicago.
Buildings sixteen to twenty stories high and people galore. Came through
Hollywood on our way here. Nice city with fine homes with pretty flowers
and big palm trees. Surely a rich man's city."

They continue to explore that city and the towns along the beach. They
attend services at a Methodist church, play pool, go to the movies, attend a
concert in the park, swim in the ocean, rent a two-room cottage in Hun-
tington Park, and find jobs in a restaurant. My father tells Marjorie: "There

is an Irishman working with us by the name of Jack Brett. He is twenty-eight years old. Has been in this country ten years but has the Irish brogue. Keeps us laughing most of the time." By the end of October they are working for the power company, putting in forms and pouring concrete. Eventually they will be climbing the high towers and receiving 95 cents an hour.

In the letters, my father often inquires about his father and the work at home. "How is pa coming with the work? Does he still go to the corner with the milk? Suppose the nights are getting chilly now. Are the potatoes dug and how are they?" Later he asks about the trading of the drakes, the gander, and the gobbler, and suggests that four hen turkeys ought to be kept. He adds, "If the Ford starts hard there is a five gallon can of winter cylinder oil upstairs in the granary. Suppose it keeps pa pretty busy doing chores now. Hope you are all well."

Seal Beach

Long Beach

At Christmastime my father receives in the mail a necktie from Kate and a shirt from Marjorie. "Many thanks to both of you for remembering me." From a girlfriend in Delavan he receives a box of six handkerchiefs, initialed. As the new year begins, the rains have come and water is up to the cottage door, and seals swim up to the beach. My father is beginning to think about buying a suitcase for when he will have to pack and return home. "The old one is pretty well shot."

The last week of February my father tells Marjorie about the boat trip that he and Mervin have taken to Catalina Island. From the deck of the *Avalon* he has taken photographs: "And if they are good will send some home. There were over 500 passengers on the boat. Saw a whale come to the top and gush water in the air three times then disappeared. After we reached Catalina Island we took a ride on the glass bottom boat. Just wonderful to see animal and plant life in the ocean. Could look down in the water to the depth of 75 feet seeing the different kinds of fish, clams and weeds." He then tells Marjorie, "Each of us bought a souvenir, an abalone shell." Sometime ago, I found the iridescent abalone shell in a box in the attic of the farmhouse. In my house, now, the abalone shell rests prominently on a shelf.

Early in March my father writes to Marjorie that he has received her letter, "and as dad says to come home will work until March 11th." He asks if the feed is holding out. And in a letter to Kate, he says that he has purchased his ticket to travel home on the Southern Pacific, will arrive in Chicago at seven o'clock in the morning, Thursday, March 26, and will probably get to Delavan on the one o'clock train. He adds: "I bought a black genuine cowhide suitcase and it won't be a bit too big for all I have to put in it."

In the days remaining, before leaving California, my dad and Mervin went to Riverside, San Diego, and Tijuana. They picked up their mail for the final time at the post office in Seal Beach. The last letter home ends: "Could plainly see the snow capped mountains from here Sunday morning which are 50 to 60 miles away. According to the papers eight inches of snow fell in the mountains Saturday, but the weather here was just like spring. No snow. Just a little shower of rain. Guess this is all for this time. Floyd."

He returned in time for spring planting and gradually took over the responsibilities of the farm. Seasons came and passed, and on a Saturday night in the fall of 1929 my father met Alice Marie Holloway at a dance in Delavan. They were married the next year in September. I was born on a day in May in 1934, and my brother was born two years later. His friend Mervin farmed for the rest of his life a few miles to the north. Occasionally he came to the house for a visit.

54

Floyd photographed by Marjorie, Delavan, 1926

In a gray metal box, I have a few letters that my father wrote to me after I left home. I read them occasionally, when I have the need to hear his voice. And I look longingly at the photographs that he took during the trip of his lifetime. This is just dandy—a word that my father often used in his letters—with me.

EMIGRANTS AND PIONEERS

My parents' marriage brought the history of other families to the farm. Joined to the families Quinney, O'Keefe, and Reynolds were the families Holloway, Bray, Wishart, and Taylor. Emigrating from England about the same time as the Irish emigrated, these families settled in the communities a few miles north of the Old Place, in the pioneering settlements of Millard and LaGrange. Through several generations, families and marriages culminated in the marriage of Alice Holloway and Floyd Quinney.

A small, spare diary remains from the ocean voyage of my mother's grandfather James Holloway to the new world. He departed from

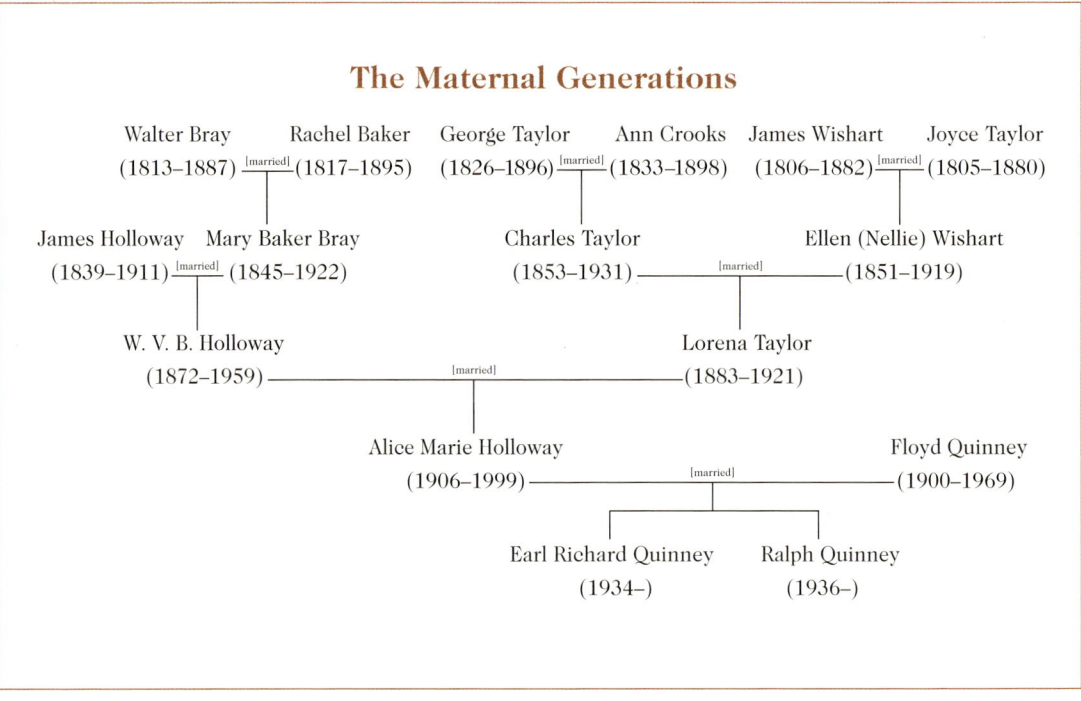

The Maternal Generations

Walter Bray (1813–1887) [married] Rachel Baker (1817–1895)   George Taylor (1826–1896) [married] Ann Crooks (1833–1898)   James Wishart (1806–1882) [married] Joyce Taylor (1805–1880)

James Holloway (1839–1911) [married] Mary Baker Bray (1845–1922)   Charles Taylor (1853–1931) [married] Ellen (Nellie) Wishart (1851–1919)

W. V. B. Holloway (1872–1959) [married] Lorena Taylor (1883–1921)

Alice Marie Holloway (1906–1999) [married] Floyd Quinney (1900–1969)

Earl Richard Quinney (1934–)   Ralph Quinney (1936–)

Liverpool on the S.S. *Damascus* in August of 1865 with the words "To America." The ship's passenger list shows his age as twenty-six and his occupation as farmer. On board he writes: "Very rough—we were very ill." "The waves keep dashing over her bow." Later in the voyage, on a Sunday: "The weather was very fine, the sea was smooth as a pond." Leaving the family farm in West Buckland in North Devon, James had purchased a ticket for passage to Australia, but changed his mind at the last minute, and booked passage to Quebec. He arrived at the port in Quebec on September 20, after the voyage of thirty days. From Canada he made his way to Walworth County, Wisconsin.

James Holloway packed a copy of the *West Buckland Year Book* when he sailed from England. I found the book in the emigration trunk stored on the front porch of the farmhouse. In the book is a poem directed to

58

the "Emigrants," to all those who will leave the old country for a life in the new world. Generations later, we seek our own good fortunes. And we dream of a paradise that cannot be found in this earthly existence. Daily we live as emigrants from another land.

Tales of virgin lands, of limitless acres that could be acquired for a song, lured the sons and daughters from the failing farms in the west of England. Soon after his arrival, James met Mary Bray, who had come from Devon in 1867 with her father and mother, Walter Bray and Rachel Baker Bray, and brothers and sisters on the ship *Nova Scotia*. They would marry, have two children, Will and Lizzie, and move several times

Walter Bray

Rachel Baker Bray

Portrait of James and Mary Bray Holloway with Lissie and Will

from one farm to another. James died in 1911, and Mary in 1922. James was remembered in an obituary as "a devoted husband, a loving father, a kind and helpful neighbor, ever ready to do a kindly act." He held offices in the county, and was regarded as a successful farmer. Will, formally known by his initials W. V. B., eventually became father to my mother, Alice Marie, and would be my grandfather.

The Taylors had started emigrating from Yorkshire in the late 1830s. Conditions in America were better than those in England for the laboring class. One of Lorena's ancestors, Joshua Taylor, Jr., already in the new world, wrote back to his brother and sisters as they were preparing to emigrate: "I am very glad you have made up your mind to come to America. I think it will be the best move you ever made in your life. It is a foolish idea that some people get into their heads there is no place like Old England, for this is a better country than England has ever been in your days or ever will be." The letter offers advice for the voyage and a list of provisions needed for starting the new life in Wisconsin.

George Taylor, my mother's maternal great-grandfather, was a cabinet-maker and carpenter who emigrated from England in 1851 with his bride, Ann Crooks. George designed and built several handsome Greek revival farmhouses that still stand in LaGrange Township. One of their sons, Charles H. Taylor, married Ellen (Nellie) Wishart in 1876. They farmed successfully on Territorial Road, and one of their children, Lorena, would become my mother's mother after marrying Will Holloway.

Ann Crooks Taylor   *George Taylor*

My mother often pointed out to me the wooded place, located on Tama-rack Road, where once stood the house that her grandmother Nellie had lived in as a child in the large household of the Wisharts. James Wishart had emigrated from England to Canada in the spring of 1828, sailing from Scarborough on the Yorkshire coast to Montreal. His parents, John and Ann Stockdale Wishart, who were of Scottish birth, soon followed him. *LaGrange Pioneers* provides a description of John and Ann: "John Wis-hart was a sturdy Scot, a mason by trade, a great reader, an intelligent talker, pious in character and temperate in his habits. His wife, Ann

Stockdale Wishart, was a woman of deep piety and noble character and like her husband was of Scottish birth and parentage." The Wisharts of Scotland descend from the Vikings of the eleventh century, and Wisharts were among the religious reformers of the sixteenth century. George Wishart was burned at the stake in 1546 at Saint Andrews.

Sailing on the same ship with James Wishart, with her family, was Joyce Taylor, who would become his wife in 1831. James and Joyce Wishart eventually settled in the village of Clinton, New York, where six of their ten children were born. In 1844 the family sailed west by way of the Great Lakes. Landing in Milwaukee, they traveled by team and wagon the forty miles to LaGrange. James had apprenticed to the trade of blacksmithing before emigrating. He continued to be a blacksmith, as well as farming the land, until he died in 1882. His wife, Joyce, had died two years earlier. Only recently have I seen a portrait of James and Joyce Wishart, sent to me by a Wishart family member in Minnesota. A grandson of James and Joyce many years ago had found and identified the mid-1850s portrait of his grandparents.

In a photograph taken on the front lawn of my mother's grandparents' house in LaGrange, a mile north of the Holloway farm, her mother, Lorena, is standing next to her mother, Nellie; grandfather Charles Taylor is sitting in a chair on the lawn, in front of the house his father built, and Lorena's brother Lloyd is sitting next to his dog. Several years ago I walked through the house with the present owner, and I imagined the many times that my mother had been in these rooms. My mother carefully labeled the photograph on the back, years after the traveling photographer had stopped to record the family on a summer's day. Photographs were to play an important part in the everyday lives of my mother's family. At the age of nine, my mother was beginning to take her own photographs, with a small Kodak camera.

James and Joyce Wishart, about 1865

Charles and Ellen (Nellie) Wishart Taylor, Lorena and Lloyd

Lorena Taylor and Will Holloway were married in the winter of 1903. My mother was born in 1906 on the farm a mile and a half north of Millard. When she was fifteen, her mother died of Bright's disease at the age of thirty-eight. Three years later her father married Mabel Stiles. My mother kept a framed portrait of her mother throughout her life. It was on the table beside her chair when she passed away in 1999, just before her ninety-third birthday.

My mother began keeping a diary in 1916, when she was nine years old, and made an entry every day for six years. With a coupon clipped from a box of cereal, my mother had ordered a camera, and it arrived on March 25, 1916. She wrote in her diary for that day: "Hunted eggs all day. I got 2 eggs for myself and 30 for mama. Played in the mud with ma's boots on. Papa went to town. I got my Kodak." On May 4th, just after her tenth birthday, she wrote: "I took my camera to school. Teacher showed me how to take a picture. Teacher took a picture of me. Jack (my cat) would not hold still so I could not take his picture. We went to Millard and back in the car. I wore my hat to school." And on May 7th she wrote: "We had a short auto ride. Took Grandma's picture. I had the head ache all day. Papa and I went to the woods. I took a picture of mama and papa in the car. Papa took a picture of me at night."

Wedding portrait of W.V.B. (Will) Holloway and Lorena Taylor Holloway

Alice Marie Holloway

The photographs mentioned in my mother's diary are in the album that she kept when she was young and growing up on the farm. I keep on the wall beside my desk the framed photograph my mother took of her mother, standing in a long dress on the lawn beside the milk house. When I look at the photographs, I know that the camera was a way of exploring the world and recording the experiences of everyday life of a family. A way of knowing that there is order in the universe, and that all might be well.

Lorena Taylor Holloway

Will and Lorena Holloway, photographed by Alice, 1916

Will Holloway, my mother's father, would live in the house in Millard for the rest of his life. For fifty-six years he held the elected position of clerk of Sugar Creek Township. In a photograph published by the *Elkhorn Enterprise* to accompany an article commemorating his many years of clerking for Sugar Creek, he is shown sitting at his desk, a plat map of the county stretched out in front of him. A rubber band is holding up the sleeve of his shirt, and a visor worn to shield the light from the lamp rests on the desk. A wooden telephone, with two bells on top and a pencil on a string dangling from the voice piece, hangs on the wall behind him.

Each afternoon, for decades, my grandfather walked to the corner store in Millard to get the evening newspaper. One February afternoon in 1959, as he walked home at dusk he was struck by a car in front of his house. He was eighty-six. Mabel lived alone in the house for several years and passed away in a nursing home in Delavan. The house in Millard now stands in disrepair as I pass it on my way to the farm. The years of visiting my grandfather and Mabel in their house and their visits to the farm as we were growing up clearly remain as part of our lives today.

W.V.B. Holloway, 1946

THE FARM

Floyd is dressed in his best clothes. Cows are grazing on the hill that slopes to the water. On the back of the photograph that Alice took of Floyd, hat in his hand, she wrote: "A happy day we spent at Whitewater Lake, June 1930. At nite we went to the show at Delavan. The nite I gave Floyd 'the answer'." Another photograph shows her standing in a park overlooking Lake Michigan. On the back of it, she wrote: "Taken in Milwaukee the day after Floyd gave me my engagement ring. Went to Kenosha, Racine and Milwaukee. Had dinner in Kenosha. Called on Kelloggs in Milwaukee and went to a show. A very happy day." This is the ring that I have placed in a box for safe keeping, along with the wedding ring that my mother gave my father the day they were married.

My mother was twenty-four when she married my father, and he was thirty. They each had had a few years of living separate adult lives before they met. He farmed the land and hauled milk to the dairy in Delavan.

Floyd at Whitewater Lake, June 1930

Alice in Milwaukee, June 1930

His album of this period shows him at work and at gatherings with friends, some surely his girlfriends. The album that she kept has photographs of parties with her friends, outings with Elma and Thelma Olsen, and an excursion to Starved Rock along the Illinois River. Late in her life she would tell about the trip to Starved Rock with her girlfriends.

Inspired, no doubt, by the houses he saw in California, my father built a bungalow-style house at the farm. It was ready to be occupied when my mother and father married in September of 1930. The wedding was held at the house of her father and stepmother in Millard. They took photographs to record the day. A framed photograph of the newlyweds stood on the dresser in their bedroom for nearly forty years. They are posing in the driveway, just before getting into the Chevrolet to begin the honeymoon that will take them on a week's trip to Niagara Falls and back through Windsor, Detroit, and Chicago. The honeymoon is documented in an album of the photographs that they took along the way.

The newly married couple settled into the new house on the farm immediately upon returning from their honeymoon. They must have kept a close watch on the Old Place, as John and Kate were in the last decade of their lives. But the photographs that follow the wedding and the honeymoon are of the new life at the farm, of the new house, and of work around the farm. With the birth of their first son in 1934, and with the birth of their second son two years later, the years from the mid-thirties to the early fifties are as fully documented as life can be with the aid of a camera.

How pleased my parents were with my birth. I am placed in all my nakedness on the kitchen table, which has been moved to the back porch

73

Wedding day, Alice and Floyd, September 15, 1930

Mabel Stiles Holloway and Will Holloway, September 15, 1930

Detroit, U.S.A., from Windsor, Ont. Canada.

Postcard from the honeymoon

Held by my father

Held by my mother

where the morning sun shines through the tall windows. On my stomach, face turned to the light, I cast an eye to the world. Throughout the year they hold me as the camera records our collective existence. Sometimes my father is in his overalls and other times he is dressed for a Sunday outing. In the snow of winter, or on a summer day, we prepare for a ride in the car. And there I am in the ubiquitous egg crate that serves as my carrier. On my first birthday, with my cake of one candle, I am placed in the south window. Never was there a baby cared for as this one.

Sometime ago I removed one of the photos from the album, and now it is in a frame on my dresser. My father has placed me on top of the straw that fills the wheelbarrow. He is pushing the load through the snow

Christmas day, 1934

My first birthday

between the milk house and the back door of the house. My mother has come out to photograph the occasion on this winter's day. On the same day, in my snowsuit, I am playing in the barnyard. Later I stand in front of the decorated Christmas tree, which has been placed on the front porch to receive the available light. What a day that was.

The album ends with the birth of my brother, Ralph. In one of the first photographs of that time, our dad is sitting on the porch holding the two of us, and I am looking worried. Next to this picture is a photo of me sitting close to the cat on the sill of the basement window. Soon I will be playfully nudging my brother as he lies on the card table for his picture in the sun. A new album will be needed to record the rest of our growing years, from the late thirties to the end of the war years.

A photograph that I have taken from the album and placed in a frame on my desk is of me at the age of three on the seat of the horse-drawn grain binder in the field north of the house. On the back of the photo, my mother has written: "Our new grain binder, 1937. Cutting grain on our very own land. How Earl loved to ride." This photograph is an early sign that I would be working long and hard days in the fields, often not to my liking. But a sign also of the attention and care that would always be given to me—on the farm, and all the time away from it.

I look at the photographs taken during the five years between 1936 and 1941, and I see the young and formative years of the two brothers, Earl and Ralph. Beyond the fascination of seeing these two boys early in life, I recognize the great attention that their parents are giving them. I count 170 photographs in the albums for this period, more photographs than for any other period in the life and times of this family. Clearly this was a time filled with new life, a time when my mother and father were most interested in recording our lives on film. With the Kodak box camera, they took turns photographing the pleasures of their daily lives. In retrospect,

With my mother and brother

On the grain binder, 1937

we who remain could weep for what has been lost—the lives of our parents, foremost. And we could, and we do, give our thanks for the good fortune of this family.

The photographs, and the making of the photographs, were not meant to show or comment on the larger social and economic conditions of the late thirties and early forties. These were the Depression years. The family farm, at least in southern Wisconsin, was still based on a subsistence economy. A team of horses drew the field equipment. Feed—oats, corn, and hay—for the small herd of milking cows was grown entirely in the fields of the farm. Other animals—chickens, sheep, and pigs—were raised mainly for use on the farm, and any surplus, especially eggs, was traded for groceries at the end of the week. At year's end, with the filing of income tax, little or no profit could be noted. Simply getting by was the order of the day, but getting by was all that was needed.

The Wild West was in our imaginations as we were growing up in Wisconsin in the 1940s. We always had horses to saddle and to ride, sometimes saddling the large workhorses. Ralph and I would listen to *The Lone Ranger* on the radio in the late afternoon, and then ride our horses, reenacting the adventures of the Lone Ranger and Tonto. On summer evenings, I rode to the end of the pasture to round up the herd of cows for milking back at the barn. Each morning, before sunrise, while doing the chores, we would turn on the radio to hear country music from Chicago's WLS. Sometimes on a clear Saturday night, we could hear the *Grand Ole Opry* from Nashville.

I thought of myself as a cowboy. Whether in play or work, or in times of trouble, I donned my cowboy clothes and faced the challenge. One

Hauling firewood

Ralph watering the Jerseys

Saturday morning when I was six, I dressed in jeans and cowboy shirt, placed a bandana around my neck, fastened a holster to my belt, and drove with my parents and brother to the Delavan Clinic to have my tonsils and adenoids removed, under ether. My lariat was always hanging on the coat rack on the back porch, ready to accompany me on the next adventure.

My cowboy life was observed and encouraged by others. I keep three postcards on my desk to this day. One card is a publicity photo of Roy Rogers sent to me by my great-uncle and great-aunt Tom and Myra Gregg during their trip to Los Angeles in 1940. Miss Roeker, my first grade teacher at Dunham School, sent me a postcard of "Greetings from Mexico" with a photo of a small boy in a Mexican hat leaning against a sycamore tree and strumming a small guitar.

Dr. Rolland Anderson—our veterinarian for cows, horses, sheep, and pigs—sent a postcard from a trip to Cheyenne, dated March 16, 1946. The card has a drawing of a cowboy herding cattle, and a poem that ends as follows:

> For a kingly crown in the noisy town
> His saddle he wouldn't change—
> No life so free as the life we see,
> Way out on the cattle range.

Dr. Anderson wrote a note to "Dear Floyd & Family" on the reverse side: "I am on my way to Colorado and waiting in Cheyenne so I thought I'd send this card to the boys. Time to go. Yours truly, Rolland Anderson." Upon his return, he brought me a fine lariat and taught me how to rope. The lariat with its brass honda hangs above my desk as I do my daily work. Mamas, let your babies grow up to be cowboys.

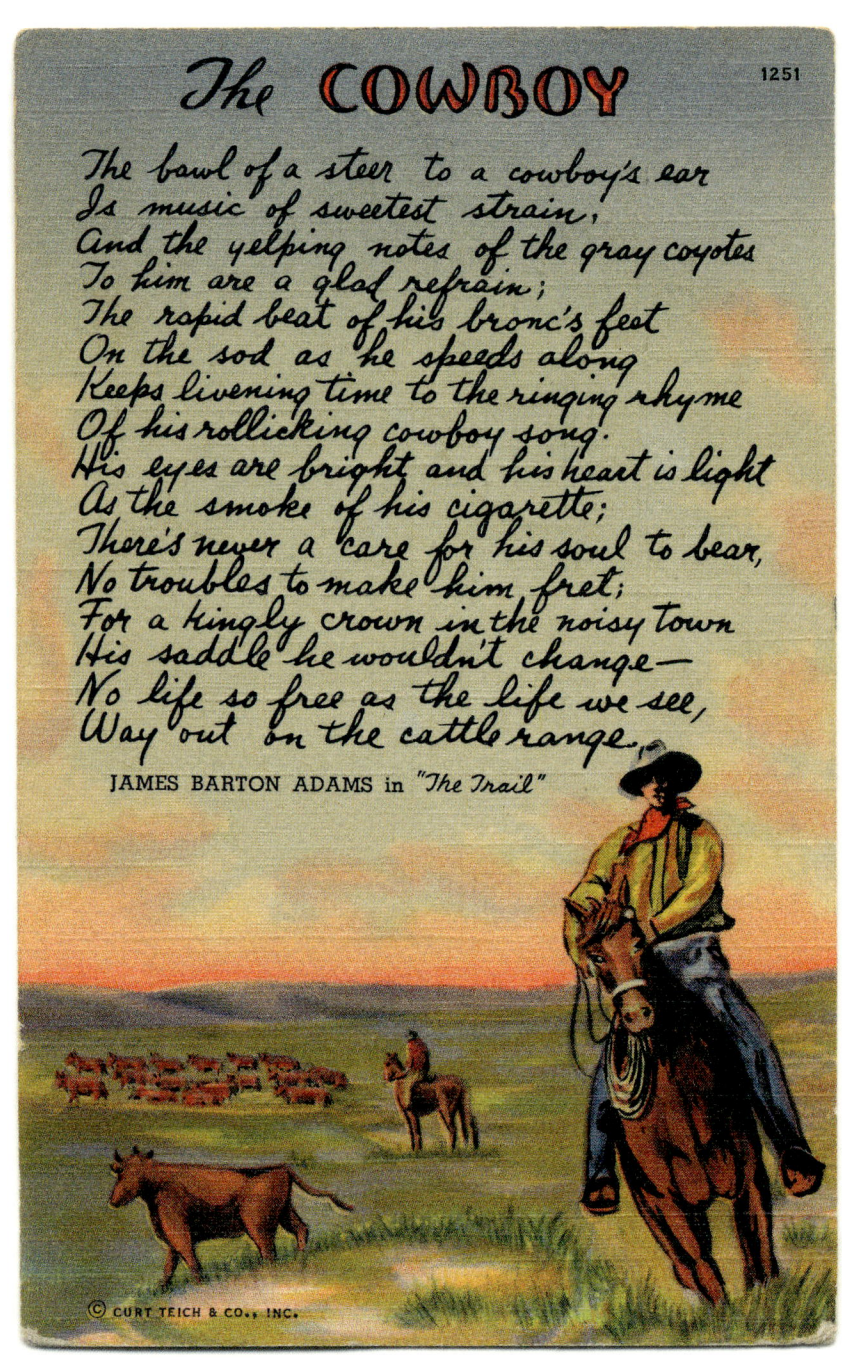

The COWBOY

1251

The bawl of a steer to a cowboy's ear
Is music of sweetest strain,
And the yelping notes of the gray coyotes
To him are a glad refrain;
The rapid beat of his bronc's feet
On the sod as he speeds along
Keeps livening time to the ringing rhyme
Of his rollicking cowboy song.
His eyes are bright and his heart is light
As the smoke of his cigarette;
There's never a care for his soul to bear,
No troubles to make him fret;
For a kingly crown in the noisy town
His saddle he wouldn't change—
No life so free as the life we see,
Way out on the cattle range.

JAMES BARTON ADAMS in "The Trail"

© CURT TEICH & CO., INC.

Cowboy postcard from Dr. Anderson, 1946

84

John Steinbeck once wrote in a letter that being at his writing desk was similar to mounting the seat of a horse-drawn wagon. Each morning he would approach his desk, climb into the seat, take hold of the reins, and ride off for a morning of travel. Since reading that letter, I have thought of myself doing the same thing each morning. My own early life of mounting the seat of the horse-drawn grain binder or the seat of the Oliver tractor for a day of work was a similar journey of the imagination. My writing—my need to write—goes back to an earlier life.

Except for the steady drone of the tractor, I would be in the silence and solitude of a full day under the big sky. There was the peace that comes with concentration on the task at hand, guiding the tractor between the rows of corn as I cultivated the land. Dream-like thoughts would come for brief times, and then disappear as I was brought back to attention. Such work has provided a model for a lifetime.

Hard work on the farm afforded a minimum of time for reading anything other than the landscape. I find a few children's books in the attic boxes, books of nursery rhymes, bird identification, and the Indian tribes of Wisconsin. Three books do stand out in my memory of childhood. One is *Hurlbut's Story of the Bible* given to me by my uncle Lloyd and aunt Elsie, Christmas of 1943. Another is *The Rover Boys at School,* a book that I would ask my mother to read to me at bedtime. The other is *The Adventures of Huckleberry Finn,* a gift from my mother and father in December of 1943 when I was nine. To this day, these three books are prominently displayed in my bookcase. They represent a literary heritage that seemed rich enough to me at the time.

My mother had taught the eight grades for five years at Bay Hill School before she was married. She had completed the nine months of rural education courses at Whitewater State Normal School. Upon marriage, her career in teaching was automatically ended by state law. I sensed that my brother and I had become her students.

My father finished the eighth grade at Dunham School, the school that I would also attend, within walking distance of the farm. Needed on the farm by his father, he completed his education by his early teens. But his love of poetry, learned and recited at school, continued throughout his life. I would be both surprised and enchanted each fall, as we picked ears of corn in the field, when he would joyfully begin to recite, "The golden-rod is yellow; the corn is turning brown; the trees in apple orchards with fruit are bending down." The intertwining of poetry and farming has tied me to the farm—as well as to poetry—for life.

I have always had the need to express myself in writing. I have also had the need to photograph, to compose and capture in an image that which is beyond the words of everyday discourse. One finds the means at hand to give substance and meaning to this daily existence.

Occasionally I will examine the photograph albums for an image of the life that once was at the farm and at the Old Place. Sometimes I write a few words about the lives of my ancestors, to remember them and to gain some understanding of where I began and who I am. And where I might be going.

A sense of wonder and purpose was gained and fostered at Dunham School, located a mile and half from the farm. The one-room building of red brick stood on an acre of land surrounded by large silver maples and American elms. Arriving at school, we would settle into our desks, placing books and tablets in the compartments under the desktops. Throughout the day, each grade, consisting of three or four students, would take its

Dunham School

Class of 1940, Dunham School

turn as a class in the front of the room. The rest of us continued with our studies, listening some of the time to our teacher, and to fellow students reciting their lessons. We learned from those who were younger, at a point where we had once been, and from those who were older and more advanced in their studies. Eight grades of students were learning together.

On the first day of school each year, the day after Labor Day, our mother without fail would take a photograph of our departure. Ralph and I would be dressed appropriately, usually in our overalls, and we would have our notebooks and lunch pails in hand or in our bags. We might be standing beside our bicycles ready to mount and take to the road that would take us the mile and a half to Dunham School. I sang joyfully, "Oh what a beautiful morning," as we pedaled out of the driveway and headed down the gravel road.

The person I would become was nurtured during the years of the war and the transformation of the family farm. At Dunham School we were given the *Weekly Reader,* and I learned about the battles being fought in countries far away. Photographs of Dwight D. Eisenhower and other heroes of the war appeared on the front page each week. Meanwhile, we collected the specimens of nature and placed them in the museum cabinet along the back wall. Radio programs of instruction in art, music, and natural history came daily from WHA, the state's educational station in Madison. After a day of play and the learning of new things, we walked home for an evening of chores.

During one of her visits to the school in 1946, the school superintendent, Ella Jacobson, summarized what learning meant for me, both at Dunham School and in subsequent years when learning became not only my work but also my happiness: "Every time I visit this school I find you doing good work. Always do your very best in everything you do. This will bring you success, and you will be happy in your work."

First day of school, September 1940

The school year would end with the annual Dunham School picnic. Before the picnic lunch began, students and adults together enjoyed a game of baseball, while the old-timers reminisced in small groups around the schoolyard. I was busy taking around the autograph book that I had started during the school year, trying to get each page filled by the end of the day. Mary presented me with an autograph that read: "You had a little lamp, very well trained no doubt. Every time your girlfriend comes, the little lamp goes out." Ending the verse, she added, "Yours till Niagara Falls." Jimmy wrote, "When you see a monkey up in a tree, just pull his tail and think of me." Our neighbor Burton Hanson wrote, "Remember me when far, far off, where the woodchucks die with the whooping cough."

We were photographed through the seasons of the year. We posed with our dogs, cats and kittens, calves, and chickens. Our pictures were taken as we worked in the fields with the horses and on the tractor. The daily farm work was tied closely to the weather and to the passage of the seasons. Winter was a time for the land to slumber under the cover of snow. There were the daily chores of milking the cows and caring for and feeding the farm animals. During the coldest part of winter, as icicles grew longer each day along the eaves of the barn, farm animals went outside only in the warmth of a sunny day.

Of all the winter chores, the morning milking was the most trying. Rising early and putting on long underwear and overalls, we left the house and walked in the darkness through snowdrifts to the barn. The cows would get up from their stalls, their breath filling the air. Cats, lazy after the night's sleep, left their beds under the straw to welcome us. The great

Holstein bull in the heavily barred stall at the end of the barn bellowed a morning greeting. Ralph and I strapped the milking machines to the first two cows, placed the cold milk cups on their teats, turned on the valve above the stanchion, and the milking began for another morning.

The small black radio tuned to WLS brought news, stock reports, and music from Chicago. On the mornings when the electricity had gone out because of a storm, we milked the cows using pressure produced by the gasoline-powered generator. If the engine would not start, we were forced to milk the entire herd by hand. We would sit on a wooden stool with our head against the cow's body to do the milking by hand.

An April breeze signaled a warming of the land. A heavy snowfall might still come well into May, but as the sun grew brighter and as geese were seen flying north in formation, we knew that winter was ending. This was the time when farmers could enjoy the pleasure of going to town every day. We ordered baby chickens at the hatchery, purchased seed corn and grain at the mill, and took plowshares to the blacksmith. I liked to accompany my father to the blacksmith's shop, where we would stand in the darkened room and watch fire and sparks fly from the forge as the blacksmith pounded a hard new edge into red-hot shares.

With spring came the time for the farm animals to give birth. By the end of March, my father would make nightly trips to the pig house. When birth was certain to take place, he would spend the entire night watching over the sows. Without tending, the old sows might roll over onto the new litter, crushing some of the young. The lambs were born without trouble, their mothers sometimes giving birth in the snow. We always felt joy in the lambing of twins. Newborn calves received special attention because of their size and their economic importance to the farm. The veterinarian, Dr. Anderson, made trips to the farm in advance of their births to detect potential problems. If he examined a cow and discovered that the calf's hooves pointed in the wrong direction, he could expect a call at his home in Elkhorn alerting him to the impending birth. To save the life of the calf and often that of the mother, he would be forced to reach into the mother and pull out the calf. Many of the calves grew to become milking cows. I took care not to become attached to the calves that were destined to be shipped each fall to the Milwaukee stockyards.

The frogs in the pond down at the Old Place heralded the arrival of spring. Their croaking and peeping sounded clearly in the evenings, as the days grew warmer. The red-winged blackbirds returned to the pond, the males perching on the tops of cattails. They scolded loudly, establishing their territory for the mating and nesting that would soon follow. Mallards returned for nesting along the edge of the pond. Tadpoles began to swim in the shallows, and grasses shot up green all around the pond. The red buds on the silver maples appeared ready to open. There were many wonders of spring in sight, but little time to linger and observe them. Spring brought more work to be done on the farm.

Some of the fields had been plowed during the fall before the snows came. The soft, moist soil in these fields was ready for tilling. The other fields needed to be plowed and disked, and cornstalks and oat stubble had to be turned. At the beginning of preparing the fields, the soil often turned to mud, sticking to the cleats on the huge rear tires of the tractor and dropping off in the driveway as the tractor was driven back from the field. Gradually the land dried in the warm sun, and dust rose as the fields were worked. Ralph and I would hurry home from school each afternoon to complete the dragging and disking of the fields. Looking over the hood of the green tractor as I pulled the drag from one end of the field to the other, I would feel small against the long horizon. I sang at the top of my voice as I steered the wheel with the palm of my hand, turning to begin another round.

With corn knives and scythe

Ralph mowing hay

Grain binder and tractor, 1944

Threshing day

We planted oats first, pulling the grain drill back and forth over the smooth fields and stopping after each round to refill the oat and fertilizer compartments. We planted corn next, and by the end of May, we watched the green tips of corn emerging from the warm soil planted with seeds only two weeks earlier. With planting completed, we would rest a bit from our labors and wait for the alfalfa, clover, and timothy to mature.

By June the hay was ready for its first cutting. The smell of freshly cut clover would spread over the fields as bees buzzed about gathering pollen. As the hay was cut, it fell into neat rows behind the mower. Sometimes this ended tragically when the harsh mower cut off the legs of a young rabbit. After allowing the hay to lie in the fields for a day or two to dry, we stacked the wagon high with hay. Inevitably, some of it dropped from the hay loader that followed the wagon across the fields. After pulling the wagon to the barn, we unloaded the hay into the mow.

Following the haying season came long days of cutting and binding the grain. The tasks for my brother and me included running the grain binder. We would hitch the McCormick-Deering binder, which had once been pulled by two horses, to the Oliver tractor. All day long, with Ralph on the binder adjusting the levers of the cutting blade and releasing the bundles, and with me on the tractor driving, we moved across the oat field. I wore a dust mask, outfitted with a penlight battery, to alleviate my hay fever, caused by the dusty grain. Each year from the time I was seven or eight years old, I complained to my father that I was being worked too hard. As I drove the tractor over rough and hilly fields, I had visions of kids in town playing and loafing while I worked in the hot sun, sneezing all day long.

At the end of the summer, the threshing machine, owned and worked cooperatively with several farm neighbors, was pulled to the field and placed south of the barn. One of my favorite photographs, taken by my mother with the Kodak box camera, shows the threshing machine,

Grain wagon, with Burton Hanson

Cultivating corn, 1947

96

powered by the old tractor turning the long, twisted belt, blowing straw into the air and into the growing stack. One man is on the horse-drawn wagon, pitching grain bundles into the hopper of the thresher; another stands atop the enormous machine, tending the threshed oats. A white leghorn hen is in the lower left-hand corner of the photograph.

Threshing time depended on the readiness of each farmer's crop of grain. When our grain was threshed, it was the responsibility of my mother to prepare and serve the noon meal to the threshing crew. Several farm wives would come to help her with the large dinner on the two or three days that the crew worked at our farm. My mother would return the help when the crew moved on to another farm. Promptly at noon, the power belt from the tractor was released and the machine fell silent. The load of grain bundles remaining on the wagon would wait until after the noon hour. The horses were placed in a shady spot and given water, and their feedbags were attached.

The metal-frame washstands, complete with Lava soap and towels, were set up in the backyard. After each thresher washed, immersing his face in the white enamel pan, the water was thrown out and fresh water was poured for the next man. The workers seated themselves on the back porch at the long oak table whose extra leaves had been added, along with additional leaves borrowed from a neighbor, to make a place for everyone. The food would arrive: mashed potatoes, meat, and gravy, followed by hot apple pie and chunks of cheddar cheese for dessert. The smoothness of the mashed potatoes established the quality of the meal. My mother would receive compliments from the well-fed crew, and some men would then wander off to finish the hour with a short nap under the Chinese elms.

The threshing season came just before the Walworth County Fair, and usually threshing was completed before the fair began. The county fair was the grand finale of the summer. It was the time to show the livestock we had been feeding and tending all during the spring and summer. We showed our calves, pigs, sheep, and chickens in the 4-H competitions. The pigs I had been raising were not ordinary pigs, but purebred Duroc hogs. By raising purebreds, all with certified registration papers, I avoided selling the pigs for pork at the end of the season. My pigs were either sold as breeding stock or retained for another year. At the fair, I would hang out a painted wooden sign to promote my Duroc hogs.

My enterprise had the objective, and ultimately the result, of providing me with the money for a college education. An ordinary fattened pig sold for about fifty dollars in the winter after nine months of slopping and feeding, but purebreds might bring three hundred and fifty dollars. Moreover, if some of my pigs won prizes at the fair, I would be assured of particularly good sales. When I finally entered college, I did not disclose to my fellow students the source of my financing.

Walworth County Fair, with the Rowleys

The unstated attraction of the county fair was being able to be away from home for several days. I spent the nights sleeping in a big tent with other 4-H members. During the day, I would roam the fairgrounds unattended and uninhibited. It was a time to greet neighbors on a new territory. My grandfather was usually in the Agriculture Building looking at the prize seed corn and vegetables. Farmers sat on and walked around the latest improvements in machinery. Neighbors picnicked on the green, listening to the band. Others viewed horse races and special acts from the grandstand. I would stop often to listen to country music being played in the tent operated by the Janesville radio station WCLO. Food was abundant: Harold Loomer's hamburgers, the Bethel Church hotdogs, Willard Olson's pronto pups, and peanut-covered ice-cream bars.

In the recesses of the imagination, there is something darkly exotic about the carnival. It was a source of magic and mystery, as well as fear. Farmers for generations kept their wives and children from the woods beyond the fairgrounds where the carnival workers encamped. They feared that the women and children would disappear along with the gypsy-like characters that they saw once a year.

For me, the carnival meant excitement, and I looked forward each year to being caught up in the sounds, vibrant colors, and crowds of the carnival. On the midway, strange-looking men and women beckoned: A woman in tight pants offering darts for popping balloons; a man with tattooed arms and an open shirt, holding out three balls to knock down a stack of wooden milk bottles. Walking past a tent with an arcade of machines, I would hear loud noises and see people wandering out with cards dispensed for a penny. Other sights would draw me on: two-headed reptiles, dwarfed men, and bearded ladies; motorcycles roaring inside a rickety-walled inverted dome; revolving wooden animals and dragons painted orange, green, and red; an octopus-shaped ride ablaze with colored lights reaching up and out into the night sky.

The county fair did not mean the end of the year's harvesting. The work, however, seemed easy compared with that of the summer, partly because Ralph and I returned to the refuge of school. Our father would spend his days cutting ripe corn, hauling it to the silo-filler, and blowing the chopped corn into the cement silo. After days of fermenting, the chopped corn turned into silage for the winter-feeding of the cows. When the best corn had completely ripened into hard kernels, it was picked by hand, ear by ear, and thrown into the waiting wagon alongside the rows. The wagon was then unloaded, the large golden ears of corn going into permanent cribs and snow fence cribs constructed to hold the overflow of the season's harvest.

Dunham School closed its doors at the end of the 1947 school year. We few remaining students were transferred to the Island School two miles west of the farm. The move to the new school and my graduation in the spring of 1948 marked the end of the old life on the farm.

The transition and adjustment to high school in town took me farther and farther away from the farm. The year was 1950, and my mother recorded the departure with box camera in hand. I had been granted a special permit for students needing to drive to school. I was in the driver's seat of the pickup truck, about to leave the farm for the beginning of my third year at Delavan High School.

Two separate worlds divided the kids in town and those in the country. We of the country were readily labeled "farmers." For the four years

Driving to high school, first day of the school year, 1950

in Delavan High School, I struggled to compensate for being from the country and to be accepted by the kids in town. After the first year, I developed severe stomach pains, and sought relief by convincing my doctor that I needed to have my appendix removed. The operation—which removed a perfectly good appendix—prepared me for the next three years of high school. I was elected to office in the student body, learned to play trombone in the band, played golf, wrote and photographed for the student newspaper, and appeared in a school play. I was invited to the house of one of the town students to watch General Douglas MacArthur on television deliver his old-soldiers-never-die speech before Congress. I was making it in town.

I graduated from high school in 1952, went to college and to graduate school, and spent the subsequent years in a series of academic jobs that would take me to places far away from the farm.

With the car headed out of the driveway, about to return to college after spending the weekend at the farm, but stuck in a drift of newly fallen snow, I shouted to my father that I was leaving this God-forsaken place and that I was never coming back. He hooked the log chain to the car and with the Oliver tractor pulled the car to the road. I knew I was on my way away from the farm. And yet, no matter how far I traveled, or how long I stayed away, I never left the farm and the farm never left me.

Even away at college, I longed to be home at night in the farmhouse. I missed the quietness at bedtime, and the soft voices bidding me good night. I worried that I would never to able to leave home. Still, I did not want to be a farmer. I wanted to move to town and to travel throughout the world. Soon I was married, and with the birth of two daughters, Laura and Anne, I had a family. I was a professor, teaching in a series of universities, and living finally in New York City. My father had often sung the song while we were working in the fields—"How 'Ya Gonna Keep 'Em Down on the Farm After They've Seen Paree?"

All the while I would return to the farm to visit my mother and father and to walk the familiar and much-loved grounds. Gradually, as my father and mother got older, they reduced the amount of work on the farm. My father passed away in the fall of 1969, two weeks after visiting us in New York City and telling us that he did not have long to live. I told him no, that he would live for a long time. My mother continued to live alone on the farm for the next thirty years. I was with her the morning she passed away, helping her prepare for an appointment with her doctor. She escaped having to leave the farm to live someplace else.

Now, I live sixty miles north of the farm. We drive to the farm and watch the fields being prepared and harvested. Ralph goes to the farm to make repairs on the buildings, and keeps the records. We all walk to the marsh and to the fields that have been planted with hardwood seedlings and with prairie grasses and forbs. Gradually the farmland is being converted to sustainable agriculture. The whole farm is becoming an integral part of the natural world. How many times have I told friends that I have spent a lifetime trying to get away from the farm? Now there is no need: I live daily knowing that the farm and I are one.

This is the story I continue to tell. I have the apparent need to keep telling it, remembering it, trying not to forget how I got from there to here. As memory of the past fades, I have the accumulated stories to help me remember. It is not that I think that the past is better than the present, or that I want to be back there again. Rather, it is by recognizing the past that I am better able to live in the present.

What I have remembered, and what I have told, is not a lament. I do miss and have great sorrow for loved ones now gone. But I also celebrate their lives. We all have our entrances and our exits. And in our time we play our many parts and live our varied lives. With good fortune, we will be remembered, just as we remember those who came before us.

THE LATER YEARS

In a sense, one never leaves home. Home is where you start from, as T. S. Eliot noted in his poem "East Coker," and home stays with you for the rest of your life.

> Home is where one starts from. As we grow older
> The world becomes stranger, the pattern more complicated
> Of dead and living. Not the intense moment
> Isolated, with no before and after,
> But a lifetime burning in every moment
> And not the lifetime of one man only
> But of old stones that cannot be deciphered.

I spent my teen and early adult years trying to extricate myself from that home on the farm in Wisconsin. I am surprised to find that after these many years I am spending much of my time and energy trying to preserve the 160 acres of farmland, woods, marsh, and the buildings that remain in various stages of decay and repair. Even among the ruins, and especially among the ruins, I find a depth of meaning. As I get older, the world becomes stranger, and more wondrous for all of that.

My home place, finally at this stage of my life, is the place I have been searching for all my life. It is the farm and the territory that radiates from it—Chicago to the south, Madison and the Wisconsin River to the north, Lake Michigan to the east, and an unlimited stretch of land to the west. This is the home country. I have reverence for the past, certainly, but I am not trying to bring back what has vanished. The present is now my home. I know that I am fortunate to be here.

Since my mother passed away, leaving the farm to my brother and me, we have been working on ways to preserve it. Several acres of hilly farmland, "the alps" north of the Old Place, have been placed in the Conservation Reserve Program. Clusters of hardwood trees have been planted on the hills. Prairie grasses have been planted on the hills and valleys—grasses commonly known as big bluestem, Canada wild rye, switch grass, little bluestem, and Indian grass. Twenty-five types of Wisconsin native forbs were planted with the grasses, including purple prairie clover, butterfly weed, compass plant, purple coneflower, goldenrod, white wild indigo, prairie dock, sky blue aster, foxglove, western sunflower, and wild garlic. Future generations will walk through fields of prairie grass and rest in the shade of the hardwood trees.

We are converting the acres of farmland to sustainable agriculture. The aim is to turn the whole farm, agricultural and otherwise, into a natural habitat. This means moving from the techniques of industrial-type farming, which relied upon the use of pesticides and herbicides, to natural and sustainable methods. The ponds and marshes, which have always been brooding areas for wildlife, will be improved for wildlife habitation. In addition, biologists and ecologists will be using the farm as a research and teaching location. The objective of all these activities is to preserve the farm, to keep it as an open space, to improve the natural habitat, and to honor our ancestors.

With the income received annually from the crops, we are able to pay the taxes and make the necessary repairs. The farmhouse is being maintained. The fate of the other buildings, including the large barn, is uncertain. Some of the buildings are already beyond repair as they begin to turn into ruins. We will let them stand, or fall as they may, as witnesses to the life that once was here. My life—and perhaps that of future generations—could not exist without the farm.

A few years ago, I photographed the things that remained on the farm. My self-imposed project was to photograph the artifacts, to make a record of the things that remain of the life that once was here. Each photograph was in the artistic tradition of the still life, where the material things of everyday life are portrayed in repose, indicating the transience of this earthly existence. These material things, devoid of their former purpose and function, received my attention and care. These old and inanimate things now have an afterlife as they rest among the ruins.

With camera and tripod, I went into the barn and climbed to the dark haymow. Stretched out on a high beam, a raccoon watched as I studied the light and made my exposures on film. Piles of hay were stacked in corners, remaining from the time the milking of the cows ceased on the farm. Below were the stanchions where cows once stood patiently. At the far end of the barn was the stall where the great Holstein bull watched our every move.

In the large metal machine shed, I photographed the artifacts from the work and play of another time. I took great care in photographing the workbench where my father last stood on the cold November day he passed away over forty years ago. A light gently fell over the workbench.

I went to what once was the chicken house, tended for years by my mother, and I photographed the tools and various other objects abandoned after they were no longer of any use. In the building that once housed a variety of animals hung the bridle and feedbag the horses once wore. The grain wagon was mired in the dirt floor of the granary.

In the basement of the farmhouse, I photographed the artifacts hanging on the walls. I photographed the sled, the milk cans, and the work caps. Canning jars once containing the preserves prepared for the family by my mother over sixty years ago lined the shelves in the fruit cellar.

With tripod and camera over my shoulder, I climbed to the attic of the farmhouse to photograph what remained while I still had the light of day. The worn winter coat that my father wore when hauling milk on winter days hung from a rafter. Scattered throughout the attic were the many things that served us well for generations.

I carefully framed in the viewfinder of my camera the furnishings in the kitchen, the dining and living rooms, bedrooms, and front porch, all in place from another time. I knew that this was the ending of an era that has stretched for nearly a hundred and fifty years on this place that I continue to call home. Am I the last of this line to bear witness to those emigrants who left the old world to escape from famine and economic depression, searching for something else?

The silo filler stored in the barn

Haymow

*Dad's
workbench
in the
machine
shed*

*Coal
scuttle,
hand wa-
ter pump,
and egg
crate
in the
chicken
house*

Horse feedbag in the chicken coop

Wagon in the granary

111

Work caps, sled, and milk cans in the base-ment of the farm-house

Canning jars in the fruit cellar

112

*The attic
of the
farmhouse*

*Ice skates
in the
attic*

113

Dad's winter coat in the attic

The living room in the farm- house

*Emigration
trunk and
rocking
chair on
the front
porch*

*Piano
stored in
the front
bedroom*

115

As I explored the past, and the remains, I thought about the writer James Agee as he once sat in a shack trying to write about the things his friend Walker Evans was photographing around him. With a spontaneous flow of words, a stream of consciousness as much as thought, Agee wrote the following in *Let Us Now Praise Famous Men:* "If I could do it, I'd do no writing at all here. It would be photographs; the rest would be fragments of cloth, bits of cotton, lumps of earth, records of speech, pieces of wool and iron, phials of odors, plates of food, and excrement." These are the artifacts, in all of their forms, from time past—from the lives that once were here. Exploring the dark corners of the old buildings, I realized that I am now one of the old ones. I am haunted by the mystery of time and place.

The phoebes had tended their nest from the early days of spring. A nest of fibers and mud clung to the corner of the porch under the eave of the farmhouse. Four feathered babies perched on the edge of the nest, about to take flight.

Phoebes are at the center of Robert Frost's poem "The Need of Being Versed in Country Things." In the poem, a farm has been abandoned, left in ruins from fire and decay. A chimney is all that remains from the house; the barn stands forsaken, the sounds of horse hooves on the floor are long gone. We humans easily lament the passing of the years and the falling to ruin of what once housed precious life. But among the phoebes there is little lament. Frost's poem ends:

> For them there was really nothing sad.
> But though they rejoiced in the nest they kept

One had to be versed in country things

Not to believe the phoebes wept.

Making my way among the ruins, I took little comfort in being versed in country things.

My mother's Bible rested on the corner of a table in the living room. Her mother and father inscribed it to her in 1917 when she was eleven. A bookmark had been placed at the beginning of Ecclesiastes. "To everything there is a season, and a time to every purpose under the heaven: a time to be born, and a time to die; a time to plant, and a time to pluck up that which is planted." The Bible was filled with newspaper clippings, church bulletins, obituaries, and pamphlets accumulated over the years. My mother saved these words near the end of her life: "We are today's seniors. A hardy bunch when you think of how our world has changed and the adjustments we had to make!"

I never asked my mother for her thoughts and beliefs about death. She once told me that she had never thought that she would live to be in her nineties. My memory is of a person who chose to live fully rather than spend time worrying about death. I am trying to do the same. Without a myth for some consolation about death, living the mystery of life continues to be my spiritual path. Death is the price we pay for living in this world. My explorations are of the wonders in the present time.

That summer of a short time ago, as with all summers, ended with the Walworth County Fair. The summer vegetables were judged and assigned ribbons. Farm animals were shown in the coliseum, and 4-H-ers gathered around the barns late into the night. The giant Ferris wheel adorned with colored lights turned against the sky. As in summers of the past, we walked through the poultry barn as we made our way out of the fairgrounds. School would begin the next morning. Much of time present, even with its spontaneous and extraordinary moments, follows a story-line.

We picked the remainder of the garden's summer crop. I completed the last photographs of the ruins and artifacts of the farm. We packed our jars of tomato sauce and salsa for the winter days ahead. I walked down to the Old Place and picked the wild grapes on the vines that draped over shrubs and trees. Back in the farmhouse, we boiled the grapes, added sugar and pectin, and poured the rich blue liquid into jars that would for a time preserve the summer's sun. As winter was about to come.

EPILOGUE

The farmhouse is empty now. The accumulated material possessions of four generations on the farm have been taken away. The attic has been cleared of the spinning wheel, the highchairs, the baby beds, the quilts, the cradle scythe, boxes of books, framed paintings and photographs, sets of dishes and silverware, and the rosary beads that served my great-grandmother Bridget for a lifetime. Taken from the basement are the tools, the sleds, the milk cans, barn jackets and caps, and the canning jars. No longer on the front porch are the emigration trunks that held the photographs, scrapbooks, farm ledgers, diaries, and the souvenirs from trips to the West. The cupboards, dressers, and closets have been emptied. Every object removed from the century and a half of living on this Midwest farm held a story and a memory.

Some of the treasures from the farmhouse will be preserved by the state and local historical societies. The old furniture and family heirlooms have been dispersed to the homes of children and grandchildren. The kitchen table that our family sat around is now in my dining room and serves as my writing table each morning. Some material objects of daily living are gone forever. Still, my desktop is covered with enough objects to last the rest of my life.

I might never have cleared the farmhouse of the material possessions of the past—the possessions of my great-grandparents, my grandparents, my mother and father, and of my generation. It could have been a museum if I had decided to make this my home in the country. But age and the desire to be at home in my house in town prompted us to clear the farmhouse and make room for others. Future renters will make their own lives in the farmhouse that I have abandoned. Of the family possessions, the upright piano—too heavy to move—is all that remains. Music will continue to fill the house.

The farm buildings, including the large barn, stand in various conditions of ruin and repair. The 160 acres of farmland, woods, and marsh continue to be maintained and preserved. Other generations will determine the course and fate of the land that has supported the previous generations of the family. Here, on yet another border, I remain, a farmer. Fortunate and happy am I for this life.

Exploration of the past informs us of the reality of impermanence. We begin to understand how our particular lives fit into the larger scheme of things, including our connection to all others in the world.

I keep reminding myself to stay focused on the present moment. Not to dwell in the past, not to remain in sorrow and regret, but to be in touch with what is happening now. I hope to dwell in the present, even as I explore how the past is shaping my life. I know that to remember the past is not to live in the past. Remembering is a vital part of living in the present. Time present is the only time we really have.

When all is said and done, after all the gaining and the spending, after trying to make sense of this life, I know that I must ultimately surrender. Everything that I cherish today I will someday be separated from. Only what I have given to others will last and not be lost.

On the cover of the scrapbook that my mother gave to me shortly after I was born is a moon looking down at five owls perched on the branches of a tree. Bats are flying in the night. There is a four-line verse in the lower corner of the scrapbook:

Five little Owls sitting on a tree,
Up rose the fair Moon, wonderful to see.
"Fly away! Dear Owls, fly away to me."
"Thank you very kindly we'll stick where we be."

Where we belong, here on Earth. Here on this farm in Wisconsin.

ANCESTOR CHART

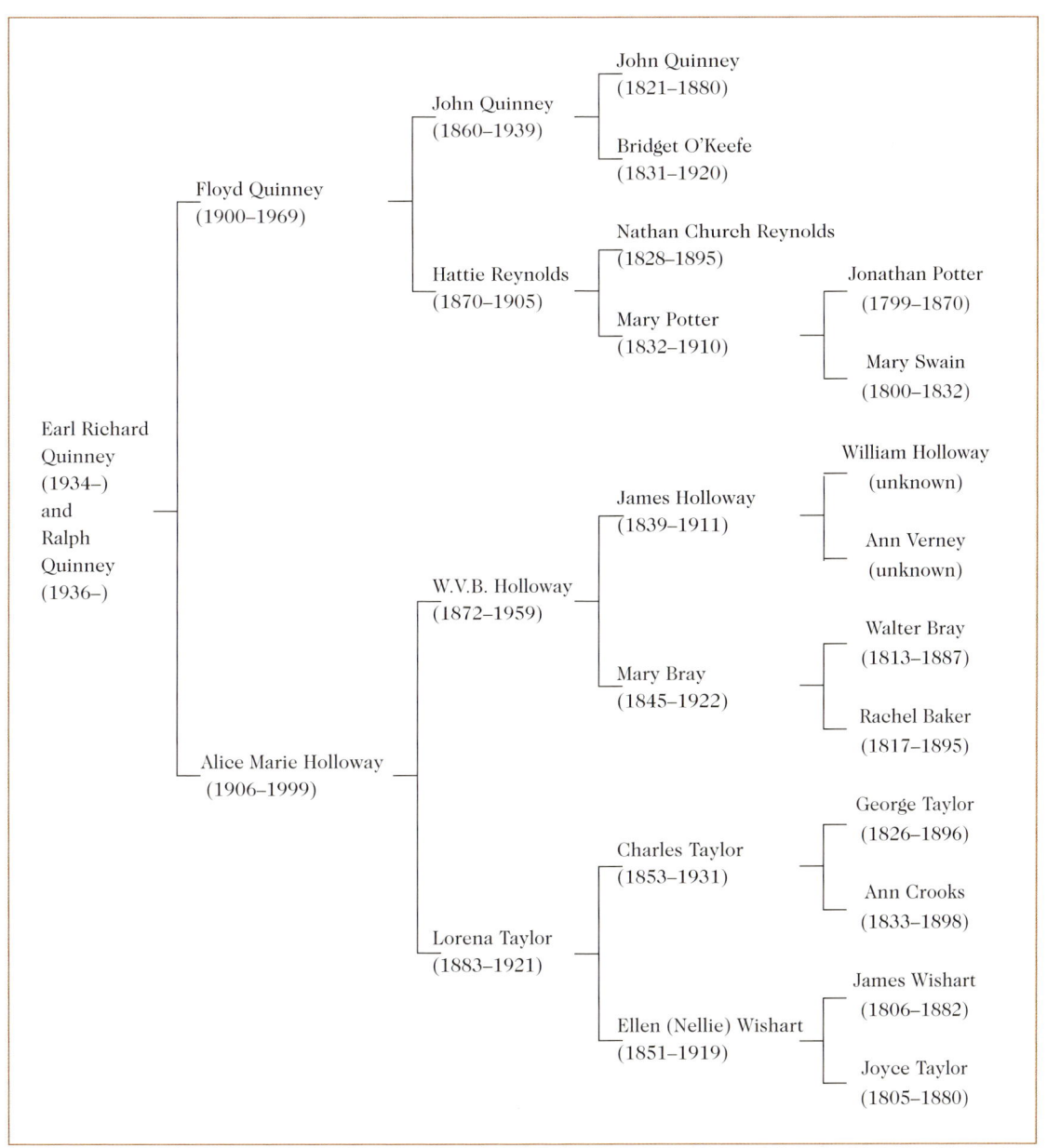

Earl Richard
Quinney
(1934–)
and
Ralph
Quinney
(1936–)

Floyd Quinney
(1900–1969)

John Quinney
(1860–1939)

John Quinney
(1821–1880)

Bridget O'Keefe
(1831–1920)

Hattie Reynolds
(1870–1905)

Nathan Church Reynolds
(1828–1895)

Mary Potter
(1832–1910)

Jonathan Potter
(1799–1870)

Mary Swain
(1800–1832)

Alice Marie Holloway
(1906–1999)

W.V.B. Holloway
(1872–1959)

James Holloway
(1839–1911)

William Holloway
(unknown)

Ann Verney
(unknown)

Mary Bray
(1845–1922)

Walter Bray
(1813–1887)

Rachel Baker
(1817–1895)

Lorena Taylor
(1883–1921)

Charles Taylor
(1853–1931)

George Taylor
(1826–1896)

Ann Crooks
(1833–1898)

Ellen (Nellie) Wishart
(1851–1919)

James Wishart
(1806–1882)

Joyce Taylor
(1805–1880)

PHOTOGRAPHS

126

BIBLIOGRAPHY

The following books provided an intellectual and aesthetic context for the writing of this book.

Agee, James, and Walker Evans. *Let Us Now Praise Famous Men.* Boston: Houghton Mifflin, 1941.

Barthes, Roland. *Camera Lucida: Reflections on Photography.* Trans. Richard Howard. New York: Hill and Wang, 1981.

Berger, John, and Jean Mohr. *A Fortunate Man.* New York: Pantheon Books, 1967.

Berry, Wendell. *A Timbered Choir: The Sabbath Poems 1979–1997.* Washington: Counterpoint, 1998.

Chaucer, Geoffrey. *The Canterbury Tales.* Trans. Nevill Coghill. Baltimore: Penguin Books, 1962.

Coleridge, Samuel Taylor. *The Rime of the Ancient Mariner.* Ed. Paul H. Fry. Boston: Bedford/St.Martin's, 1999.

Delano, Jack. *Photographic Memories.* Washington, DC: Smithsonian Institution Press, 1997.

Donnelly, James S., Jr. *The Great Irish Potato Famine.* Gloucestershire, England: Sutton Publishing, 2001.

Eliot, T. S. *The Complete Poems and Plays of T. S. Eliot.* London: Faber and Faber, 1969.

Frost, Robert. *Collected Poems, Prose, and Plays.* New York: Library of America, 1995.

Garland, Hamlin. *Main-Travelled Roads.* New York: Macmillan, 1891.

Garland, Hamlin. *A Son of the Middle Border.* Lincoln: University of Nebraska Press, 1979. [1917]

Heaney, Seamus. *Finders Keepers: Selected Prose, 1971–2001.* New York: Farrar, Straus and Giroux, 2002.

Hirsch, Marianne. *Family Frames: Photography, Narrative, and Postmemory.* Cambridge: Harvard University Press, 1997.

Jackson, J. B. *The Necessity of Ruins.* Amherst: University of Massachusetts Press, 1980.

Kaplan, Laura Duhan. *Family Pictures: A Philosopher Explores the Familiar.* Chicago: Open Court, 1998.

LaGrange Pioneers. Walworth County, WI: LaGrange Ladies' Aid Society, 1935.

Leopold, Aldo. *A Sand County Almanac.* New York: Oxford University Press, 1949.

Lesy, Michael. *Wisconsin Death Trip.* New York: Pantheon Books, 1973.

Liebling, Jerome, Christopher Benfry, Polly Longsworth, and Barton St. Armand. *The Dickinsons of Amherst.* Hanover, NH: University Press of New England, 2001.

Logan, Ben. *The Land Remembers.* New York: Viking Press, 1975.

Logsdon, Gene. *The Contrary Farmer.* White River Junction, VT: Chelsea Green Publishing Co., 1994.

Masters, Edgar Lee. *Spoon River Anthology.* Ed. John E. Hallwas. Urbana: University of Illinois Press, 1992. [1915]

Naipaul, V. S. *Literary Occasions: Essays.* Intro. and ed. Pankaj Mishra. New York: Alfred A. Knopf, 2003.

Quinney, Richard. *Borderland: A Midwest Journal.* Madison: University of Wisconsin Press, 2001.

Quinney, Richard. *Journey to a Far Place.* Philadelphia: Temple University Press, 1991.

Quinney, Richard. *Where Yet the Sweet Birds Sing.* Madison, WI: Borderland Books, 2006.

Sontag, Susan. *On Photography.* New York: Farrar, Straus and Giroux, 1977.

Stegner, Wallace. *Where the Bluebird Sings to the Lemonade Springs.* New York: Random House, 1992.

Steinbeck, John. *Steinbeck: A Life in Letters.* Ed. Elaine Steinbeck and Robert Wallsten. New York: Viking Press, 1975.

Sudek, Josef. *Josef Sudek, Poet of Prague: A Photographer's Life.* Biographical Profile by Anna Farova. New York: Aperture, 1990.

Suzuki, Shunryu. *Zen Mind, Beginner's Mind.* New York: Weatherhill, 1970.

Travis, David. *Edward Weston: The Last Years in Carmel.* Chicago: Art Institute of Chicago, 2001.

Vendler, Helen. *The Art of Shakespeare's Sonnets.* Cambridge: Harvard University Press, 1997.

Wang, Wei. *Laughing Lost in the Mountains: Poems of Wang Wei.* Trans. Tony Barnstone, Willis Barnstone, and Xu Haixin. Hanover, NH: University Press of New England, 1991.

Wescott, Glenway. *Good-Bye Wisconsin.* New York: Harper, 1928.

Wolfe, Linnie Marsh. *Son of the Wilderness: The Life of John Muir.* New York: Alfred A. Knopf, 1945.

Wordsworth, William. *The Poems of William Wordsworth.* Ed. Jonathan Wordsworth. Cambridge, England: University Printing House, 1973.

About the Author

Richard Quinney is author of several books that combine autobiographical writing and photography, including *Journey to a Far Place, For the Time Being, Borderland, Once Again the Wonder, Where Yet the Sweet Birds Sing, Of Time and Place, Field Notes, A Lifetime Burning,* and *Once Upon an Island.* His other books are in the field of sociology. His retrospective book of photographs, *Things Once Seen,* received the August Derleth Award from the Council for Wisconsin Writers. He lives in Madison, Wisconsin.